WEATHER PERMITTING

Twenty-Five Years of Ice Storms, Hurricanes, Wildfires, and Extreme Climate Change in Canada

CHRIS ST. CLAIR

Published by Simon & Schuster Canada

New York   London   Toronto   Sydney   New Delhi

SIMON &
SCHUSTER
CANADA

A Division of Simon & Schuster, Inc.
166 King Street East, Suite 300
Toronto, Ontario M5A 1J3

This Simon & Schuster Canada edition November 2022

SIMON & SCHUSTER CANADA and colophon are trademarks of Simon & Schuster, Inc.

For information about special discounts for bulk purchases, please contact Simon & Schuster Special Sales at 1-800-268-3216 or CustomerService@simonandschuster.ca.

Manufactured in the United States of America

Interior Design by Wendy Blum

10 9 8 7 6 5 4 3 2 1

Library and Archives Canada Cataloguing in Publication

Title: Weather permitting : twenty-five years of ice storms, hurricanes, wildfires, and extreme climate change in Canada / Chris St. Clair.
Names: St. Clair, Chris, author.
Identifiers: Canadiana (print) 20220211906 | Canadiana (ebook) 20220212066 | ISBN 9781668002889 (softcover) | ISBN 9781668002896 (ebook)
Subjects: LCSH: Canada—Climate. | LCSH: Severe storms—Canada. | LCSH: Natural disasters—Canada. | LCSH: Climatic changes—Canada.
Classification: LCC QC985.S234 2022 | DDC 551.6571—dc23

ISBN 978-1-6680-0288-9
ISBN 978-1-6680-0289-6 (ebook)

The whiskered face was drawn and tired. A half-smoked cigarette was tucked behind his ear. He looked older than his fifty years. The clothes he wore were oil stained and dirty and hung loosely on his frame. He was stooped forward. His calloused hand was outstretched; it held a chocolate bar.

The youngster's father stood beside her and nodded his approval. She took the gift.

The old man rose to his feet and faced the father. Both men had tears in their eyes as they slowly shook hands and patted each other on the back.

Thank you.

The old man would provide free gas, a kind word, help, and comfort to hundreds of evacuees from Fort McMurray that weekend.

This book is for you, and anyone else who has helped or been helped in time of need.

CONTENTS

"First Nations peoples have a special relationship with the earth and all living things on it. This relationship is based on a profound spiritual connection to Mother Earth that guided Indigenous peoples to practice reverence, humility, and reciprocity. It is also based on the subsistence needs and values extending back thousands of years. Hunting, gathering, and fishing to secure food for self, family, elderly, widows, the community, and for ceremonial purposes. Everything is taken and used with the understanding that we take only what we need and we must use great care and be aware of how we take and how much of it so that future generations will not be put in peril."

ASSEMBLY OF FIRST NATIONS

WEATHER PERMITTING

INTRODUCTION

"Hi, I'm glad you're here" is how I would begin almost all my broadcasts as Canada's weatherman. For more than twenty-five years, I was a host on The Weather Network and reported on the country's major weather events for the CBC, too. The job took me from coast to coast, surveying snowstorms, ice storms, hurricanes, and heatwaves. It also gave me the privilege of speaking to Canadians and asking them how they were shaped by these often brief but massive disruptions to the environment and climate. For some, these events were life-changing.

My fascination with the weather began as a child living in Nova Scotia, on the shores of Bedford Basin. The long arm of Halifax Harbour reaches inland, and Bedford Basin is a large bay, 8 kilometres long and 5 kilometres wide. Hills rise from three sides of the basin; Rockingham and its hillside homes lie on the west shore. On many days the Basin would disappear into the fog and the foghorns would blare, their sound and echo dulled in the mist. On certain evenings the fog would literally roll up our street, and my friends and I would ride our bikes in and out of its shroud. Fog is, hands down, my favourite type of weather.

During the summer and fall, my family often took a drive to Peggy's Cove. We'd climb over the rocks and explore the area, always watching the ocean. We knew to play only on the light-coloured rocks and to never go near the dark-coloured ones, which were slippery and wet. The warning plaques hadn't been erected then, but those who lived by the sea knew where not to venture.

In the fall, the air would cool, and sometimes there were strong storms when the rain would come in torrents and the wind would bring down tree branches. The strongest seemed to arrive around Thanksgiving, and would remove the last colourful leaves. I remember the occasional power outage, but hurricanes and tropical storms were less frequent then. But I was in Peggy's Cove in 1971 when Hurricane Beth hit the province. My sisters and I were standing in our raincoats eating cookies, and through the sweeping wind and rain we watched the massive swells burst against the rocky shore beyond the lighthouse.

Winters on the east coast were always long and varied. Snow, sometimes a lot of it, was usually followed by a thaw and rain. We struggled to keep a skating rink in our backyard and a snow fort by the driveway. The frequent winter thaws meant that sledding and skiing were intermittent activities, but they brought my friend, fog, and the eerie nights when the street lamps cast a diffuse orange glow on the city.

I attended Duc d'Anville Elementary School, and it was there that my love for history and geography began. The school was named for the French explorer whose fleet of forty-four vessels hobbled into Bedford Basin in 1746 while on an expedition to recapture Louisbourg and Acadia from the British.

Thousands of crew were sick with scurvy, typhus, and typhoid, diseases they would spread to the local Mi'kmaq and Acadian communities. Most of the expedition, including the Duc d'Anville, perished and were buried near Birch Cove, not far from where I lived. I knew those forests and shoreline—I played there as a child and wondered what it must have been like back then.

In grade four, we were taught about the Halifax Explosion, the largest human-made explosion before the atomic bombing of Hiroshima in 1945. On December 6, 1917, at the height of the First World War, the SS *Mont-Blanc*, a French cargo ship laden with explosives, collided with the SS *Imo* from Norway, obliterating the north end of Halifax and leaving 9,000 injured and 2,000 dead. At the time only 65,000 people lived in Halifax, and a significant portion of the population was made instantly homeless. To further compound the tragedy, in the hours following the blast, a massive blizzard struck the Maritimes and Nova Scotia particularly hard. So interesting to me then was how the hilly terrain deflected the initial shockwave from much of the southern half of the city, which saved so many lives. History and geography.

At Halifax West High School, I had three teachers who fuelled my love for these subjects. It is from Mr. Hersey, Ms. Buren, and Ms. McBurney that I learned that Halifax's Mi'kmaq name is K'jipuktuk, which is pronounced *che-book-took*, or Chebucto to a Haligonian. The name means "great harbour," which attests to the city's importance as a seaport. During the great wars, ships assembled by the hundreds before embarking across the Atlantic to Europe.

It was during this time that I also got a job doing a

radio show. All my life I enjoyed listening to the radio. I remember dialling in to WCBS in New York to hear Walter Cronkite's captivating report on the 1968 *Apollo 8* moon mission. For the first time ever, humankind had actually guided a space shuttle to the moon and come back safely— amazing stuff at a time when space travel was still in its infancy. The historical impact of that event and the power of Cronkite's words made that Christmas unforgettable. Howie Meeker's play-by-play of NHL hockey games was just as enthralling—I could almost smell the popcorn and feel the chill of the rink air.

So with these inspirations in mind, I called David Wolfe, the program director at CJCH radio in Halifax. It was 1978 when David met with me and offered me a couple of weeks of training. I caught on quickly, and soon enough I was hosting the all-night show on Saturdays and Sundays.

Throughout the 1980s, I worked at several radio stations in eastern Canada, and I eventually became a program director. That job was about bringing talented people together to create shows that were funny and relevant, and building the profile of the hosts so that they became community influencers. It was a thoroughly enjoyable undertaking, developing the distinct personality and sound of a radio station.

I had so many wonderful mentors. Brian Phillips, the legendary morning man at CJCH, taught me about humour— where to find it and how, when, and where to employ it. Great radio programmers and talent developers like Terry Williams guided me in the art of brevity and word selection, both of which are important elements in building compelling stories.

By the early 1990s I was the program manager at a radio

station in Montreal. But the work had lost its appeal. I was no longer feeling challenged. So I did what so many people do when they are in their early thirties and at a crossroads: I decided to learn how to fly airplanes.

I have always been fascinated with aircraft, the principles of flight, and navigation. I learned about carburetors, engine ice, air law, instrument flight, and meteorology. About a third of what you learn while attaining your pilot's licence is the weather. Countless hours are spent studying how the atmosphere behaves, why it behaves the way it does, and what variables it presents as obstacles to safe flight. To me it was incredibly interesting, and I further indulged myself with dozens of books on meteorology.

There is joy in piloting an airplane—looking at the geography from above, seeing glacial drumlins and drainage patterns . . . always keeping an eye open for a suitable place to make an emergency landing. The process of piloting is structured in systems to ensure safety and keep the workflow between flying the aircraft and communicating with controllers smooth and unencumbered with clutter.

I loved the process of gathering information and then applying it to my flight plan. Before a flight, I would go to the Environment Canada office at the airport to get a briefing on the current and forecast conditions. The airport meteorologist and I would sometimes spend an hour talking about the variances and subtleties of the forecast.

I decided that flying airplanes was going to be my new career, so I got busy building up time in the cockpit and collecting ratings on my licence to fly at night, fly on instruments, to fly bigger, faster multi-engine planes. It was eating through a lot of cash. At the same time the aviation industry was contracting, and jobs for commercial pilots weren't as plentiful as they had been a year or two earlier.

That's when a radio colleague in Montreal suggested that I contact The Weather Network. She thought my knowledge of weather, coupled with my broadcast background, might work well for them.

At the time, the majority of the on-air weather presenters were not meteorologists. Instead, a collection of skilled television hosts gave the forecast. There were some exceptions, but we hadn't quite entered the age when a specialist was required to lend credibility to a story. The Weather Network and its French counterpart, MétéoMédia, did, though, employ a full staff of seasoned meteorologists, who explained the story to the presenters, who in turn reported on air to the viewers.

I began working at The Weather Network on Saturday, October 28, 1995, the weekend prior to the last referendum on sovereignty and independence in Quebec. On my drive from my home in Kingston to Montreal that day, hundreds of "Stay" banners hung from overpasses along the highway. There was a feeling of unease and hope in the country, and I knew that feeling was everywhere in Canada. On Monday the people of Quebec narrowly chose to remain a part of Canada. There was a national sigh of relief.

My thoughts about Canada that weekend formed the basis of how I would always present the weather—as a story that we all share individually and collectively. The land, the sea, and the climate make us who we are. We cannot change the weather. We simply thrive in or endure it. When the elements turn against us, we come together to protect our communities and lend a helping hand.

To make my presentations interesting, I decided that I

would be the voice for the weather, explaining and describing what was happening in the atmosphere and the environment and why it was happening. The forecast would become short science classes, so that viewers could understand *why* it was going to rain, snow, or hail.

One of my favourite mottos that The Weather Network used was "Before, During, and After the Storm" because it captures the dedication to the work that we journalists and meteorologists have. We want to prepare those who could be impacted by the coming conditions. For the rest, who follow along and enjoy the pastime of watching weather, we want to illustrate the impact of the event and tell the stories of the communities, families, and individuals who experienced and witnessed firsthand nature's awesome display.

In Canada, weather unites us, no matter how it impacts us. We are proud to say we live in the city that has the most snow in any given winter or the highest temperature ever recorded over the summer months. This book is about the stories behind the storms we have survived over the past twenty-five years. It underscores who we are as a people, and I hope it casts a light on important changes that are impacting us now and will do so into the future.

WEATHER PERMITTING

CHAPTER 1

Red River

Winnipeg, Manitoba, 1997

Ten thousand years ago, a lake once covered much of what is now Manitoba, Saskatchewan, the Great Lakes region, and the northern states of Minnesota and North Dakota.

Lake Agassiz formed at the end of the last ice age, during the Wisconsin Glacial Episode, thirty thousand years ago. Estimates are that the ice sheet was 13 million square kilometres in size and held 25 million cubic kilometres of ice. The weight and motion of the ice left a depression in Earth's surface. Then, over thousands of years, the massive glacier began to melt and retreat northward.

The ice age was ending. Our planet was well into another warming cycle. The northern Arctic Ocean, with its southern arm, Hudson Bay, was thawing and providing open water during the summer months. The warmer salt water was eating away at the northeastern corner of the ice sheet that had covered North America for ten thousand years. That ice sheet had lowered sea levels and provided the land bridge that allowed for human migration from Asia to the American continents.

Over eight thousand years ago, the ridge of ice that separated the lake from the open water of Hudson Bay suffered a fracture. In an instant, millions of tons of ice collapsed into the bay. The sound would have been deafening.

A tsunami followed the cascade of ice. The metres-high wave rushed across Hudson Bay, radiating its energy into the ocean. Lake Agassiz poured through the ever-widening rift; the dynamic force of so much moving water would continue to tear away at the lake's northeastern rim. Trillions of cubic metres of freshwater flowed into Hudson Bay and, ultimately, the world's oceans. Indigenous stories of a great rising of the sea and even the biblical flood may have been a description of this event. It's estimated that ocean levels around the world would have risen by 1 to 3 metres.

The addition of all that freshwater into the Arctic Ocean decreased the ocean's salinity, altering the density of the seawater. The resulting change in the ocean currents had a direct impact on global weather patterns, which affected agriculture and human migration.

Lake Agassiz must have been an amazing sight. At its peak, Agassiz was larger than any lake on Earth in our time. The northern and eastern shores met massive walls of ice that rose into the sky. Waterfalls as high as Niagara poured meltwater from the glacier, filling the lake. When the last full retreat of ice began, water drained eastward and southward, through a valley called the Traverse Gap into the Glacial River Warren, which emptied into what is now the Mississippi River system and the Great Lakes and St. Lawrence River.

Lake Agassiz takes its name from the nineteenth-century Swiss-born geologist and naturalist Louis Agassiz, who studied glaciers and the theory of ice ages and their impacts on nature and human civilization. The very first scientific studies of what is now Manitoba were conducted by the American geologist William Keating in the 1820s. Keating found compelling evidence that a massive lake once covered much of the area. Fifty years later, the American geologist and archaeologist Warren Upham confirmed that Keating was correct—a huge prehistoric lake had once covered Manitoba. Upham named the now vanished lake in honour of Agassiz.

All that remains of this once massive inland sea is Lake Manitoba, Lake Winnipeg, Lake Winnipegosis, and drainage routes into Hudson Bay, the Great Lakes, and the Gulf of Mexico. What was once the floor of Lake Agassiz is now rich and fertile farmland. Strange ridges, escarpments, and gravel berms cross the landscape. These geological oddities are evidence of the ancient past and also shape the present drainage patterns of this vast lowland.

One such ridge lies along the Minnesota–North Dakota border. Rivers flow on both sides of the ridge; on one side they feed the Mississippi and all its tributaries, and on the other, the Red River and its tributaries.

The Red River flows northward through the lowlands of North Dakota and into the flat plains of southern Manitoba, meeting the Assiniboine River in Winnipeg. Winnipeg is the Cree word for "muddy water." Long before the first Europeans arrived in 1738, Indigenous people had used the location as a meeting place for trade. The fur trade led to the establishment of a colony in the early 1800s and to the Fort Garry settlement. Indigenous people who visited the colony told settlers stories of great floods in the region.

In 1826, a massive flooding of the Red River destroyed Fort Garry and the colony. The water flow is estimated to have been 6,400 metres per second. The landscape was altered forever, and the lives of everyone in the Red River community were severely disrupted.

When the water receded, Fort Garry was rebuilt on slightly higher ground, at the forks of the Red and Assiniboine rivers, in what is today the heart of Winnipeg. The settlement was renamed Upper Fort Garry.

The area would flood again in 1852, 1861, 1892, and 1897. In fact, the Red River would flood with increasing frequency as Winnipeg grew in size and importance as an agricultural trading centre, transportation hub, and seat of government.

Ancient Lake Agassiz left both a gift and a predicament. The gift is productive soil for farmers—some of the most productive in North America. But beneath the rich soil is a thick layer of clay that prevents good drainage. Heavy rain or a deep, rapidly melting snowpack can quickly saturate the ground with moisture and lead to overland floods.

The other problem is that the Red River flows from south to north and is the main drainage system for thousands of square kilometres of flatland. The south warms faster than the north, so as the winter snowpack melts and fills the Red River in the south, the water flows north into areas that are very often still frozen.

The lowlands of southern Manitoba are a natural flood-plain prone to annual spring floods. Weather extremes will occasionally magnify the scale and scope of flooding. The behaviour of both climate and weather in the year leading up to Winnipeg's Great Flood of 1950 brought together the perfect array of elements for an epic environmental event.

Conditions that contributed to the deluge began the previous autumn. A new record for rainfall was set in October 1949; as winter set in, the ground was saturated with moisture.

Then in December, January, and February, the province got more snow than it had over the previous fifty years. March was bitterly cold, and the deep snowpack was slow to thaw that spring.

By April, the Red River was melting in North Dakota. A high volume of water began flowing north toward still-frozen Manitoba, and ice jams formed near the Canadian border. Compounding the meltwater was a cold rain that fell nearly every day that month. As the rain fell, the snow melted, but the ground remained frozen, and a massive lake began forming in the southern part of the province. The lake grew by 40 square kilometres per day, and by late April, it measured 1,000 square kilometres. Farmland and villages in the Red River Valley were submerged. There were fears that Winnipeg would flood too. The Red Cross, the military, and thousands of volunteers began filling sandbags so that a series of dikes could be built around the city.

On Friday, May 5, cold arctic air raced east across the prairies and hit a surge of moist and mild air from the Gulf of Mexico. Strong wind and torrential rain fell across Manitoba— 34 millimetres that afternoon—then heavy snow.

That night, eight of the levees ruptured. The relentless, strong wind created huge waves that crashed against the dikes, and cold water from the massive lake gushed into several Winnipeg neighbourhoods. The rapid movement of flood-water freed some of the ice jams that were clogging the Red River south of the city. A flash flood pushed the swollen river over the banks; it raced northward into the city and washed away four of the eleven bridges.

Manitoba premier Douglas Campbell called on Prime Minister Louis St. Laurent for help and declared a state of emergency. Water was flowing through Winnipeg at 3,000 cubic metres per second. The lake south of the city measured 65 by 100 kilometres.

Over the next several days, dozens of Winnipeg neighbourhoods were evacuated. At the time, it was the largest evacuation in Canadian history—100,000 people, or 34 percent of the city's population, were forced to leave their homes.

The water rose for two more weeks.

On May 26, the Red River flood peaked at 10 metres above normal levels. Ten thousand homes had been damaged or destroyed. It would take nearly three weeks for the water to recede.

What could be done to prevent the next disaster?

Duff Roblin, who would become the premier of Manitoba in 1958, advocated for a floodway.

Construction of the Red River Floodway, or "Duff's Ditch," began in 1962. Over 76 million cubic metres of earth was moved to build the 47-kilometre floodway, as well as a series of dikes, levees, and control gates. At the time, it was the second-largest earth-moving project on the planet; only the Panama Canal was bigger. The floodway can move floodwater around Winnipeg at 2,500 cubic metres per second. The project came in ahead of schedule and under budget, having taken six years to complete at a cost of $63 million.

In late April 1997, I was showing an enhanced satellite image of southern Manitoba on The Weather Network. The shimmer of water—a shining, smooth, black surface—covered what should have been a patchwork of fields. "This lake is now 2,000 square kilometres in size, almost twice the size of the lake that formed in 1950," I said. "The floodwater will peak in Winnipeg by the weekend, but even after the peak, the water will take weeks to slowly recede. Let's hope the dikes and sandbags hold in Winnipeg."

The conditions that led to the Great Flood of 1997 were almost identical to those of 1950—an unusually wet autumn, a snowier-than-average winter, a cold spring, then a rapid thaw accompanied by a strong storm with heavy snow and powerful winds.

On April 17, Grand Forks, South Dakota—a city with a population of more than fifty thousand—was evacuated when the Red River burst through the banks. Water rose a metre higher than the dike that had been built to protect the city. Then a fire erupted downtown, destroying dozens of buildings and apartment blocks.

In southern Manitoba, hundreds of farms stood like miniature islands, surrounded by sandbags heaped into piles 8 to 10 metres high to hold back the sea of water that now covered the fields.

Sandbags, dikes, and levees held back the great lake from the small towns and villages in the Red River Valley too. But in Ste. Agathe, 20 kilometres south of Winnipeg, the dikes failed and the village was overtaken by the lake.

This was far worse than the Great Flood of 1950.

On April 23, Winnipeg mayor Susan Thompson declared a state of emergency and ordered the first in a series of mandatory evacuations. In South Winnipeg, 9,000 residents would have to leave their homes. South of Winnipeg, 25,000 people had been evacuated as the water continued to rise, and 8,000 were living in shelters in the capital.

The military had deployed 8,500 troops to build levees and dikes and to fill sandbags. An army of 70,000 volunteers worked at their sides. More than 8 million sandbags were delivered to neighbourhoods across the city, and more than 45,000 truckloads of clay were used to build temporary dikes.

Students volunteered to fill white plastic bags with sand, the results of which were jokingly referred to as "Red River perogies." Neighbours helped each other stack walls of sand-bags around their homes. Commuters on the ring road, which itself was a levee, could see front-end loaders moving tons of sand, rock, and earth into position.

The Brunkild Dike was erected during the last three weeks of April. Built of boulders, mud, and crushed rock, the dike formed an 8-metre-high levee that stretched 40 kilometres from Brunkild to St. Norbert, a suburban neighbourhood on the west bank of the Red River. The dike was constructed to keep the ever-growing lake from flooding Winnipeg from the southwest.

As in 1950, a storm rolled across Manitoba, bringing heavy rain and strong winds. There was concern that the waves lap-ping at the Brunkild Dike would collapse it. Soldiers installed erosion-resistant plastic sheets along its entire 40-kilometre length as gale-force winds roared and a cold rain fell. Aside from a few leaks, the levee held back the water.

On Thursday, May 1, the Red River peaked as it moved through Winnipeg; the water was 8 metres above the level of winter ice and flowed just a few centimetres beneath the roadbeds of the river's bridges. Massive chunks of ice banged against the structures as they passed below, the noise reso-nating dully amid the sound of churning water. Damage to the Provencher Bridge, at The Forks, would take months to repair.

The Red River Floodway was diverting 2,000 cubic metres of water per second away from Winnipeg; the river itself was carrying 2,100 cubic metres of water per second on its path through the city. Without the floodway diverting half the volume of water, the river would have submerged downtown Winnipeg.

Premier Gary Filmon, a hydraulic engineer, said, "We've still got another week of white-knuckle time." The prolonged water pressure on the temporary sandbag dikes could cause them to leak or fail. Dozens of crews—municipal, military, and volunteer—worked around the clock to survey and perform maintenance on the over 200 kilometres of primary dikes that surrounded Winnipeg.

Over the next three weeks, the massive lake slowly shrank, draining into the Red River. On May 23, Mayor Thompson declared the emergency over. The floodway had done its job.

Once again, Winnipeg had been spared devastation, but it was apparent that more work needed to be done to mitigate the impact of the annual spring floods. Engineers used extensive computer modelling to help guide their flood mitigation projects. The new protection system was designed to withstand a once-in-seven-hundred-years flood.

Over several years, beginning in 1999, the Red River Floodway in Winnipeg was enlarged. Twenty-one million cubic metres of earth was excavated to expand the floodway's capacity; rather than 2,500 cubic metres per second, it could now carry over 4,000. The soil that was removed to expand the floodway was used to build new dikes and reinforce the existing flood protection system. In addition, a series of reservoirs were built to hold excess runoff during the peak stages of future floods.

Ten years later, the Red River Floodway would be tested again. In March 2009, the Red River flooded in North Dakota.

Governor John Hoeven declared a statewide disaster as the river rose 12 metres above flood stage. Then a snowstorm struck, adding more volume to the river and across the flooded fields.

In Grand Forks, the Red River rose 16 metres, inundating the city with water. Manitoba was still covered in a blanket of snow and ice, and the water poured overland. A giant lake formed again.

Highway 75, Manitoba's main route to the United States, was closed due to the floodwater. Farther north, the Red River Floodway couldn't be opened because the river was still frozen and ice jams could damage the infrastructure and potentially flood the city of Selkirk, just north of Winnipeg.

By April, massive ice jams were clogging the Red River and its tributaries north of Winnipeg. Several communities were evacuated because of flash floods. Forecasters predicted the flooding would be nearly as bad as the 1997 disaster. The Province of Manitoba deployed Amphibex icebreakers to the Red River and the floodway. The machines look like floating tanks with several backhoe shovels mounted like huge steel arms. These arms are used to break thick ice and pull apart ice jams to get the water flowing.

By April 8, the ice had been cleared and the Red River Floodway was opened. Water cascaded through the channel and around Winnipeg. The Red River also flowed through the city. The water crested at nearly 7 metres, a metre below the level recorded during the 1997 flood. Without the expanded floodway, it's estimated that the crest would have been at least 2 metres higher downtown, over 9 metres in total.

There is no ultimate solution for flooding in southern Manitoba. It is flat land, a floodplain, filled with ancient meandering rivers. Human innovation has resulted in the Red River Floodway and the series of levees and dikes that protect Winnipeg and many other communities; however, the weather will always be a wild card.

This area is an ideal location for storms to mature. The topography and geography permit a confluence of warm moist and cold arctic air masses. That merging results in powerful snowstorms in the winter and torrential rain in the spring and autumn.

"Winterpeg, Manisnowba" is the snidely cute nickname that tells only part of the story of the city whose motto is "Unum Cum Virtute Multorum," or "One with the strength of many."

Manitobans and Winnipeggers have always faced extreme weather head on. They have found ways to mitigate the impacts of the annual spring floods, and they get their fair share.

In June 2007, the most powerful tornado in Canadian history touched down in Elie, 45 kilometres west of Winnipeg. The winds were estimated to be nearly 500 kilometres per hour. The very next month, on July 25, just 65 kilometres southwest of Winnipeg in Carman, Canada recorded its highest ever humidex reading at 52.6 degrees Celsius.

CHAPTER 2

Ice Storm

Quebec and Ontario, 1998

It was sunny and 5 degrees Celsius in Montreal on the first Sunday morning of 1998. Icicles dripped steadily outside our brightly lit studios, and I could feel the warmth of the sun as it beamed through the window. On my desk was a series of maps that recapped how much snow had fallen during the Christmas and New Year's holidays—over 30 centimetres in both Ottawa and Montreal.

I was formulating how to explain the rollercoaster weather pattern we were experiencing in eastern Canada—the biting cold and heavy snow of Christmas, then the thaw over the past weekend, to be followed by a very complex storm that was due on Monday. The lead-up was easy; the coming storm system was vexing.

In the atmosphere above eastern North America, a warm, moisture-laden air mass was coming in from the Gulf of Mexico and moving along the spine of the Appalachian Mountains toward Ontario, Quebec, the Maritimes, and much of New England.

At the same time, much colder air was beginning to circulate southward around a high-pressure ridge located over the far north of Quebec and Labrador. Cold air is dense; its molecules are packed closely together. A cubic unit of cold air will weigh more than the same cubic unit of warm air, whose molecules are much more widely spread. The weight of air is a confusing concept. A teacher of mine had a good analogy: in cold air the molecules are huddled together to try to keep each other warm, while in warm air they stay away from each other because it's too hot.

The weather maps I looked at were the classic models for an ice storm.

The last time a big ice storm had hit Montreal was in February 1961. Freezing rain fell for two days, resulting in the accumulation of nearly 40 millimetres of ice. Compounding the accretion of ice were winds of over 140 kilometres per hour. Trees fell onto power lines, and thousands of people were left without electricity for several days. The extensive property damage led to changes in the city's infrastructure, so that Montreal would be more resilient during future ice storms.

Freezing rain is the most misunderstood type of weather event. Freezing rain is not ice, falling like rain from the sky. Ice that falls from the sky is sleet or graupel or ice pellets, to be meteorologically correct. Freezing rain is far more insidious.

Freezing rain occurs when the temperature of the shallow layer of air at the surface is below the freezing mark. When rain, from a pool of warmer air aloft, falls into this colder, shallow layer, it instantly freezes on contact with any below-zero surface. This creates an interesting illusion: though it appears to be raining, each supercooled raindrop coats the surface in ice. Vehicles, roads, buildings, trees, power lines—everything is covered in ice.

The longer the rain continues, the greater the accretion of ice. This process doesn't end until the moisture flow is exhausted, the surface temperature rises above freezing, or the air aloft cools uniformly so that its moisture content can freeze to ice pellets or snow.

During a typical winter, many regions of Canada will experience freezing rain. However, most episodes last for several hours—a duration long enough to make it look pretty outside, but not so long as to cause significant issues.

Patrick de Bellefeuille, my counterpart on our French-language service MétéoMédia, and I discussed the possibility that this might become a long-duration storm, perhaps even a record-setting one.

"The cold air will fill all the valleys and stay in place until this pattern breaks down," Patrick said to me. "Look on the map—there is another high-pressure area near Bermuda that's going to keep the cold air over the east. This is going to last all week. It's bad, really bad."

By the time I finished my broadcast at lunchtime on Sunday, January 4, there was growing concern about the potential severity of this storm. Environment Canada had issued warnings for freezing rain.

Driving home to Kingston, some 300 kilometres west, I noted that the skies had become overcast and the bases of the clouds were sinking lower. The temperature was dropping, too.

On Monday morning it was raining. The dense, cold air had not yet circulated this far west, but in both Ottawa and Montreal the temperature at ground level had fallen below zero. Just a thousand metres above the surface, the temperature was several degrees *above* zero. This was the moisture flow from the Gulf of Mexico.

From eastern Ontario through southern Quebec, rain was falling into the shallow cold layer at the surface and creating freezing rain. The temperature profile in the atmosphere was more uniform near the Atlantic Ocean and offered upward of 30 centimetres of wet snow.

By Tuesday morning, Montreal and Ottawa had already experienced nearly twenty-four hours of continuous freezing rain; at my home in Kingston, it had been over eighteen hours. Trees had begun to sag and bend under the weight of the ice. Roads were slick, covered in a gleaming layer that made both driving and walking hazardous. School had been cancelled; businesses were opting to stay closed. The power was beginning to go out across the region.

From Kingston in the west, north to Ottawa, south to Upstate New York, and eastward through the St. Lawrence River Valley to Montreal and the Eastern Townships of Quebec, ice was coating every surface. Road travel was treacherous, and de-icing crews were unable to keep up with the accretion. People were waiting out the storm in their homes; only essential services remained open for the public.

Throughout Tuesday the rate of rainfall increased dramatically, and the ice built up rapidly on all surfaces. In Kingston, the trees were shedding icicles, endlessly clinking when they hit the ground, like the toasting of a hundred wineglasses. By evening the tinkling of broken icicles was accompanied by what sounded like shotgun blasts, two or three every minute. Tree limbs and entire trees were buckling under the weight of the ice. Outside, the night sky had an eerie green aura from the glow of chemicals burning in the hundreds of electrical transformers that had short-circuited or exploded.

Farther east, in Quebec, not only were trees collapsing but power lines and poles had begun snapping from the strain of the ever-growing ice. By late evening large areas of southern Quebec and eastern Ontario began steadily losing electricity. It was as if someone was simply flipping a switch and turning off the lights, one town at a time.

The cold quickly overtook the stored warmth in homes. Inside, the temperature dropped, from 18 to 15 to 10 to 5 degrees Celsius and colder. Heat came from fireplaces . . . if you had one.

That evening I took a walk through my neighbourhood. The temperature was –2 degrees Celsius. Rain was falling steadily and freezing onto my coat. The road, the sidewalk, the lawn—everything was one continuous sheet of ice. I slid like a kid down the street, then shuffled slowly along the sidewalk after I slipped and fell on the cold, hard ice.

We had power on my side of the street, but two blocks away the electricity had gone out. Live power lines lay across backyards. The orange lights of a public utilities truck blinked as two lineworkers inspected the wires. Walking on this wet and icy surface was doubly dangerous—the downed power lines now added the threat of electrocution.

"This street, this block, and others will be off grid," a lineworker told me. "The lines are coming down too fast for us to keep up. If more come down, stay away from them."

As I slowly shuffled my way back home, I saw extension cords running across the street, from one house, down the driveway, across the road, over a snowbank, and into a neighbour's home. People were sharing anything they could, including electricity.

As we lay in bed at night, I could hear the rafters and joists in my house groan and creak from the weight of the ice that was accumulating on the roof.

That night the first surge of moisture in the atmosphere began to ebb, and the steady rain became a persistent freezing drizzle. Closer to the surface, the temperature continued to fall as much colder air from the north surged into southern Quebec and eastern Ontario.

On Wednesday, January 7, more power lines failed under the weight of ice or fallen trees, and several electrical substations shorted out. By the afternoon, thousands of people in both provinces had been without power or heat for over forty-eight hours. Emergency centers began opening to take in those who needed shelter.

Communication was becoming increasingly difficult. Cellphone towers were without electricity or had collapsed under the weight of the ice, power failures disrupted landline telephone service, and radio-transmitting towers lost service, cutting off television as well as AM and FM broadcasts. It was going dark, and the forecast anticipated a new push of heavy freezing rain for Thursday and Friday.

I telephoned The Weather Network to discuss the storm and determine if I should attempt to make the trip to Montreal on Friday. The Téléport building, which housed our studios, was using diesel-powered generators, and the electricity was being rationed for only essential lights and electronics. The heat had been turned off. There was enough fuel to run the generators for three days, then the building would go dark.

Our conversation about the ice storm was bleak. A blast of much colder air was being directed southward. The temperature would drop a further 10 to 15 degrees Celsius, and that cold would last for at least a week. None of the accumulated ice would melt in the coming days. In addition, more moisture was set to arrive on Thursday, in the form of freezing rain.

"Don't come to Montreal—it's not safe," I was told.

My next call was to Ted Silver, the program director at Q92 and CIQC 600 radio in Montreal. I did a weekend radio show on his stations and wanted to get his take on travelling to Montreal.

"As I drove along the 20 this morning," he said, "the transformers were exploding like bombs beside the highway. It's really bad."

The stations and their transmitters were running on gas generators and, like at The Weather Network, the fuel supply was limited. "We've got to decide which station to shut down. There's no power to pump fuel and we're going to run out in the next day or so," Ted said. "It's like a disaster movie here."

Ted went on to explain that the radio and television stations were pooling resources so that they could continue to broadcast emergency information to the public. Only three English stations were on the air, two AM and one FM. Along with the French broadcasters, they had one mission: to advise and assist the public.

Premier Lucien Bouchard had declared a state of emergency and requested the federal government send in the military for assistance. Sixteen thousand troops were sent to Quebec and Ontario in the country's largest peacetime military deployment.

I decided to stay in Kingston.

Of the four radio stations in Kingston, only one remained on air. It was on the same underground power line that served

Hôtel Dieu Hospital. My friend Jim Elyot was the morning host on one of the stations.

"Our transmitting tower, the tallest one in eastern Ontario, just crumpled and fell," Jim said. "Our engineers are trying to rig up a temporary tower on the side of the building, but it's so icy and really dangerous. It will be another day or two before we can broadcast."

The story was the same in Ottawa, Brockville, Cornwall, across southern Quebec, and in northern New Hampshire, Vermont, and New York. The power was out, it was getting colder, more freezing rain was falling, and communicating to the public was becoming nearly impossible.

The final surge of moisture arrived on Thursday, January 8. More rain fell into the increasingly cold pool of air at the surface, resulting in more freezing rain. Ice was now more than 100 millimetres, or 5 inches, thick. The weight it was imposing on structures far exceeded their design limits.

It must have been unworldly to witness the cascading collapse of the massive hydro towers that supplied southern Quebec, eastern Ontario, and the northeastern United States with electricity. The noise would have been deafening as one after another fell. Tons of ice and steel buckled, screeched, and fell onto the frozen fields and highways. Twisted metal and thick high-voltage power lines lay across the landscape in the Montreal area. In all, over 1,000 steel towers fell and 35,000 wooden power poles snapped, leaving almost 5 million people without electricity. When viewed from space, there was a huge black hole in the massive patchwork of lights across eastern North America. Nearly 20 percent of Canada's workforce did not make it to work that day.

By Friday morning, after eighty hours, the freezing rain finally ended.

We were lucky at my house. We had power for most of the ice storm. In total we lost electricity for only eighteen hours because the power lines were underground. Our only interruptions came when transformers and substations encountered problems. To this day I am thankful for the design forethought that is an important part of mitigating climate impacts on our infrastructure.

Millions were not as fortunate.

On the farms of eastern Ontario and western Quebec, an agricultural disaster was underway. Hundreds of barns and animal sheds had collapsed under the weight of ice. Beneath the flattened structures were dead livestock; thousands had frozen to death due to lack of heat, and more had died when the buildings collapsed. Pig, chicken, cattle, and dairy farms lay decimated; 300,000 animals died as a result of the storm. Dairy production in Canada fell by 25 percent, and a million litres of milk that was being processed was destroyed when the power failed.

Millions of trees had died or been damaged by the ice, including much of the sugar bush in Quebec, severely impacting the world's largest maple syrup industry. Fruit orchards in both provinces lay in ruin. It would take twenty to thirty years for those trees to regenerate. At Mount Royal Park in the heart of Montreal, 5,000 trees were so irreparably damaged they had to be cut down. In fact, 80 percent of the 140,000 trees in the park were damaged by the ice.

In Canada, 28 people had died from hypothermia, carbon monoxide poisoning, or heart attacks. In the U.S., 16 people died; 945 suffered heart attacks, carbon monoxide poisoning, or broken limbs from falls.

Over the next few days, the power slowly began to return,

one street or neighbourhood at a time. Chainsaws buzzed as municipal crews worked tirelessly to remove felled and damaged trees. The icy trees and branches were cut up and stacked beside snowbanks to await the giant woodchippers that slowly moved through communities.

Power crews from across Canada and the United States worked around the clock beside the exhausted local teams. Together they erected new power poles, replaced burned-out transformers, and strung new lines, connecting homes and businesses to electricity again.

In the larger urban areas, power was being restored in good time. In rural areas it would take much longer—two, three, even four weeks. The reconstruction of large transmission towers would take months to complete, and sporadic power outages continued into late February.

The skies were clear and brilliant blue on Monday, January 12. It had been a week since the ice storm had begun. Now high pressure from the Arctic had moved over eastern Canada, bringing with it the coldest air of the new year. The temperature struggled to climb above –10 degrees Celsius. The brilliant sunshine gleamed off the icy surfaces—it appeared as if we were living in a world made of glass. It was time for me to head into Montreal to The Weather Network studio.

Driving east on Highway 401 was slow going. The roads were coated in ice and rutted with deep grooves where cars and trucks had driven during the storm. On the four-lane highway, only one lane was usable in each direction. Abandoned cars, trucks, and semi-trailers were in ditches on both sides; some were tipped on their sides in the centre median. Over the 90 kilometres between Kingston and Brockville, I counted nearly a hundred abandoned vehicles.

This part of the drive took almost two hours on the still slick highway. Sand and salt had begun melting the ice, but it would take days for that effort to have an impact.

As I headed alongside the St. Lawrence River, I was stunned by the impact of the ice damage. It looked like a giant had stood over eastern Ontario and swung a massive scythe, removing the top half of all the trees. Those that remained unbroken were bent over, nearly horizontal to the ground.

All of this destruction was covered in a sparkling layer of ice. The landscape was both spectacular and majestic in its beauty and horrific in the magnitude of damage to the woodlands. There are only estimates of the number of trees lost to the ice storm, but they are in the millions. Studies of the forest canopy after the ice storm indicated that more than half of all trees incurred severe damage. Several species, including the silver maple, saw significant damage to over 90 percent of all trees.

I stopped for gas at a service centre near Cornwall, knowing that there would be shortages of both fuel and electricity in Montreal. After I filled my car, I walked across the icy parking lot and into the restaurant to get a coffee and a sandwich. There weren't many customers, just a couple of truckers and a waitress at the counter. I stood behind the truck drivers and asked them about their drive.

"I'm heading back to Montreal with a load of bottled water," one told me. "The roads are not good, but a bit better than when I drove out of Montreal this morning. The power is coming back too, but the south shore is still dark."

"The service centre just opened up this morning," the waitress said. "Where I live the power went out on Tuesday and came back on Sunday. You really take having electricity for granted, especially after it goes off for so long."

I wished them a good evening, gathered my coffee and sandwich, and walked back into the cold air to finish up the drive to Montreal. Darkness was descending as I pulled back onto the highway. The setting sun cast a red hue on the glistening ground. The headlights of my car flashed over the icy trees, which sparkled like diamonds.

My usual three-hour drive was now up to nearly six hours, and the condition of the highway had steadily worsened. I was making 40 kilometres an hour and had lost count of the number of abandoned vehicles in the ditches. The familiar warm lights of Coteau-du-Lac were absent; so were the lights at the autoroute interchange near Dorian. The exurbs of Montreal were struggling to be rewired to the hydro grid; for the people who had stayed in their homes, this would be their eighth night without electricity.

I made my way to downtown Montreal, toward my apartment. Turning onto my street, I was heartened to see light glowing through the windows of the three-storey apartment buildings that lined the road. There was power and there would be heat.

On Tuesday, January 13, my alarm woke me at 3 a.m. I washed up, had a quick bowl of oatmeal, and headed into work. Inside the Téléport building the power was on. It had been restored over the weekend. They had run on the diesel generator for five days.

"The smell of diesel was sickening. We all had headaches," Debra Arbec, our morning anchor, told me.

We compared stories about the storm as we planned our continuing television coverage. Patrick de Bellefeuille would join us from the streets of Montreal via a remote camera. The ice storm would be one of the first major weather events that the network would broadcast live outdoors.

The morning was bright, sunny, and cold. People were beginning to make their way to the high-rise, office-towered canyons of Montreal. A new danger had developed: large sheets of thick ice were falling from buildings. Some fell straight to the sidewalk and shattered into thousands of pieces. Others slid from the sides of buildings and were lifted by the wind. The five-inch-thick slabs of ice would tumble and break into pieces the size of a table a block or so away. The falling ice had forced the closure of sidewalks and streets in the central business district. It was simply too dangerous to walk or drive in much of downtown. The quaint streets of Old Montreal were also closed to pedestrians as icicles rained down from eaves and the sides of centuries-old buildings.

A quick thaw, with temperatures above zero, would melt all the ice in just a few days; but the temperature would not rise above freezing until February 1.

The cost of the 1998 ice storm was massive, in terms of lives lost, losses to the agriculture and forestry sectors, and damage to critical infrastructure. Estimates place the cost at over $5 billion in Canada and the United States.

The storm was also an awakening for many people. Once-in-a-lifetime storms or weather events were now happening with greater frequency; they were also becoming more severe. It was a glaring signal that our climate was changing and that the impact would be costly in damages to infrastructure and to lives.

CHAPTER 3

The Two Juans

Nova Scotia, 2003 and 2004

After the 1998 ice storm it was clear to me: climate change was having an impact on the severity and the frequency of large-scale weather events. These are the hall-marks of rapid global warming. Unfortunately, in climate science and meteorology making a clear and definitive link between singular occurrences and the larger threat to human-ity is not a simple task.

In the spring of 1998, The Weather Network relocated its operations to Toronto. At the time I co-hosted *The Morning Report* and *Good Morning Toronto*. I also began working on a series of short features that explained how various aspects of the weather worked, and how some weather events in particu-lar were signs that our climate was changing far more rapidly than the public imagined. It was called "The Science Behind the Weather," and it is still in production and available for viewing on multiple platforms.

Tricia Bell was the producer and director, while I wrote and hosted the series. We would explain various phenom-ena, like fog, freezing rain, tornadoes, snow squalls, the UV index, and humidex values. Together we travelled to various locations in all types of weather to record the vignettes. Then, in post-production, we added special effects and animations to deftly make complex science easy to understand. Often an impending or ongoing storm offered us opportunities to create new episodes that would help explain weather terms that we used regularly during our broadcasts but that are unfamiliar and confusing to people outside of meteorology.

Hurricane Juan offered us such an opportunity. That deadly storm allowed us to produce segments that helped explain what a storm surge is, how the Saffir–Simpson Hurricane Wind Scale is measured, and how hurricanes get their names.

Hurricane Juan formed on September 24, 2003, southeast of Bermuda. It was the tenth named storm of the year and followed the Gulf Stream to Atlantic Canada. The temperature of the seawater was 3 degrees above average.

Heat is the energy that strengthens a storm. By Saturday, September 27, Juan was racing toward Nova Scotia. I was working on The Weather Network that morning, explaining how hurricane strength is categorized by wind speed. Tropical storms have winds between 63 and 118 kilometres per hour. Category 1 hurricanes have winds between 119 and 153 kilometres per hour. The wind in Category 2 is 154 to 177; in Category 3, 178 to 208; Category 4, 209 to 251; and anything over 252 kilometres per hour is a Category 5 hurricane. Juan now had winds of 165 kilometres per hour and was a Category 2 hurricane on the Saffir–Simpson Scale. I recited the scale and its numbers as a satellite image of Juan played behind me on television.

Is that ever boring to watch, I thought after the segment.

It was raining outside our studios when I got the idea of

how to better present the information to the public. Tricia and I discussed how we could shoot and produce a couple of "Science Behind the Weather" segments about hurricanes that afternoon.

"We need material to help explain how hurricanes behave," I said.

"Where do you want to film?" Tricia asked.

"Let's drive to St. Lawrence Park in Port Credit. It's on the water, and if we shoot with the lake behind us, it'll look like we're at the ocean," I said.

We recorded an explanation of the Saffir–Simpson Hurricane Wind Scale. It was raining and the waves were splashing against the dock. Stronger winds would have made for a more dramatic scene, but it still looked good.

"How about if you stand in the water beside a dock to explain storm surge," Tricia suggested.

I waded into Lake Ontario and stood waist deep in the cold water.

"Storm surge is an upwelling of water caused by the force of a powerful and constant wind," I detailed. "It leads to flooding and can cause immense damage along shorelines."

We added text and video of hurricane damage to the segments and put them on air to augment our coverage of Juan. The storm was drawing closer to landfall, now just over twenty-four hours away.

Shelley Steeves, our reporter in Halifax, had recorded interviews with Nova Scotians as they prepared for the arrival of Juan that Sunday night. They were getting fuel for gas generators, stocking up on groceries, and stowing away lawn furniture.

The radar I played that day showed the outer bands of rain

along the south shore. "It's going to be the wind with Juan," I stressed, "because the trees are still in full foliage. They will catch the wind and come down. The power is going to go out."

Juan made landfall at Shad Bay, near Peggy's Cove, just after midnight. The eye of the storm was 35 kilometres wide, and its northern eyewall passed directly over the city of Halifax at one in the morning on Monday, September 29.

Landfall coincided with high tide, a worst-case scenario for storm surge. The powerful wind literally lifted and drove the sea onto land. Buoys at the mouth of Halifax Harbour snapped their moorings after measuring 20-metre waves. A 2-metre rise in sea level was recorded at Historic Properties, where two schooners sank. Massive boulders the size of refrigerators that lined the seawall were lifted by the waves onto piers and the promenade.

Dozens of boats sank at yacht clubs in the Halifax area, and dozens of containers were blown off two ships moored at the South End Container Terminal. In Dartmouth, several freight train cars were washed into the harbour, and the main rail line was severed when erosion washed away the railbed along the shore of Bedford Basin.

The wind was measured at 180 kilometres per hour on an oil rig in Halifax Harbour. The airport recorded its strongest wind gust ever, at 143 kilometres per hour, though unofficial estimates suggest it might have been as high as 230 kilometres per hour. Anemometers, the instruments that measure wind speed, were torn apart by the wind both in the city and at Sambro Lighthouse; in previous storms they had measured winds up to 193 kilometres per hour.

The wind damaged 31 percent of all homes in the metro area. Victoria General Hospital, the largest medical centre in Atlantic Canada, was evacuated when the roof began leaking. Several high-rise residential buildings in the city were evacuated because of the high winds, and travel along the two suspension bridges that cross the harbour was halted.

On the morning of Monday, September 29, almost 900,000 people were without electricity. For thousands, it would be two weeks before the power was restored.

Tree loss was extensive. In Point Pleasant Park, the beautiful 75-hectare woodland lost nearly 75 percent of its trees. A Department of Lands and Forests survey concluded that millions of trees were brought down in Nova Scotia and Prince Edward Island.

Eight lives were lost to the storm. The damage was estimated at $300 million. Juan was the most devastating storm to strike Halifax since 1893.

Considering the damage and deaths, the Canadian Hurricane Centre asked, for the first time ever, that the World Meteorological Organization retire a hurricane name. The request was granted.

Then, just four and a half months later, another powerful storm struck Nova Scotia, so powerful it would be dubbed "White Juan."

White Juan developed in the warm Gulf Stream off the coast of Cape Hatteras, North Carolina, on Tuesday, February 17, 2004. The storm was a nor'easter: one formed as cold continental air from eastern Canada flowed over the warmer water of the Atlantic Ocean and Gulf Stream. The contrasting temperatures quickly intensified the storm—its central barometric pressure fell 37 millibars in twenty-four hours.

The upper-level winds of the jet stream also helped to invigorate the storm. By Wednesday, February 18, it was travelling northeast toward Nova Scotia at 40 kilometres per hour. "Winter Storm Warning," "Blizzard Warning," and "Wind Warning" flashed on television screens that afternoon. The snow was already falling in Nova Scotia, and visibility had dropped rapidly.

The storm began slowing down, unusual behaviour for a nor'easter, which would usually accelerate as it moves north. In the case of White Juan, the surface low pressure centre aligned with an upper atmospheric low, which both intensified the strength of the storm and slowed down its forward motion.

This storm would now pass slowly along the coast of Nova Scotia and then across eastern Newfoundland.

In a storm of this nature, the heaviest snow accumulates to the west, or left on a map, of the storm's track. Typically in a nor'easter there is upward of 20 or 30 centimetres of snowfall, but because this storm was moving slowly, the estimates were greater than 50 centimetres. Some computer models even suggested 70 centimetres of snow.

As video showed cars struggling to climb the hilly, snow-covered streets of Halifax and children sledding down Citadel Hill, we drew attention to another feature of this storm.

"Because it is moving at just 27 kilometres per hour, the swath of heavy snow has spread in a wider path. Almost all of Nova Scotia, PEI, and southern New Brunswick are going to be blanketed with a half-metre," I said. "The wind is also a serious issue today. The gusts are currently at 80 to 120 kilometres per hour. This is the worst snowstorm in decades."

On Thursday, February 19, Shelley Steeves sent a report explaining that snowplows could no longer keep up with accumulation. Snow was coming down at a rate of 3 to 5 centimetres per hour as night fell. The wind in Halifax was steady at 80 kilometres per hour with a gust recorded at 124. Visibility was less than 200 metres and often down to zero. The plows came off the roads. Being outdoors was no longer safe.

On Prince Edward Island, the Special Olympics National Winter Games were suspended as storm surge flooded coastal areas during high tide. In Pictou, Nova Scotia, the water reportedly rose more than a metre; wind- and wave-driven ice floes ate away at the shoreline along the Northumberland Strait and Gulf of St. Lawrence.

On Friday, February 20, the storm was east of Newfoundland. Slowly, throughout the day, conditions began to improve in the Maritimes. The strong northerly wind eased, steady snow tapered to flurries, and by late afternoon the sun was peeking out from behind the clouds. The air was filled with the sound of shovels biting into the deep snow.

John Hamm, the premier of Nova Scotia, declared a state of emergency and ordered a 10 p.m. curfew so that roads could be cleared of snow. The curfew remained in place throughout the weekend as provincial, municipal, and private snow-removal crews fought against the 2- to 3-metre snowdrifts that covered highways and city streets. Many side streets and residential areas would not see a snowplow until Monday. The impassable snow-clogged roads also led to delays in power restoration for thousands of customers.

Getting an accurate measure of exactly how much snow fell was difficult because the wind had created so much drifting. Officially, the city of Halifax recorded 95.5 centimetres of snow, Yarmouth received 101 centimetres, and Charlottetown tallied nearly 75 centimetres.

Radar images suggested that convective snow, or thundersnow, may have produced snowfall rates of nearly 20 centimetres per hour in some locations and that the true accumulation might have been as high as 150 centimetres. Thundersnow is a thunderstorm in winter. It is rare, but it does happen when there is a great clash in temperatures within a turbulent air mass. As with thunderstorms, great volumes of moisture are lifted high into the atmosphere, and the electrical charges created by the movement of so many water particles—and in the case of thundersnow, ice particles—produce the thunder and lightning. Just like a thunderstorm with its torrential downpours of rain, thundersnow offers heavy bouts of rapidly accumulating snow.

White Juan would produce the largest snowfall in the city's history. Weather records dating from 1872 show that the amount of snow that fell was more than 40 centimetres higher than any previous storm. The cleanup continued for days. The 2-metre-high snowbanks that lined the streets would last well into March. In Halifax, tens of thousands of tons of snow were dumped into the harbour because there was literally nowhere else to pile it. Financially, the two storms left a multi-million-dollar trail of damage in Halifax and throughout the Maritimes.

Atlantic Canada is not a stranger to vicious winter storms and hurricanes; the weather history here is filled with stories of gales, blizzards, ice storms, and floods. The often beastly weather has driven ships on shore, capsized oil rigs, and resulted in countless tragic deaths. But that same weather has provided a universal binding spirit in the communities of Atlantic Canada, where the people and the weather are one.

CHAPTER 4

High Water

New Brunswick, 2008

B ig, wet snowflakes painted the night sky opaque. It was February 2008. Families were circling the rink in front of Moncton City Hall, and a young boy imagined he was Sidney Crosby, stickhandling a hockey puck across the ice behind Shelley Steeves, our reporter in New Brunswick. Shelley was filing a story about the most recent in a series of heavy snowfalls that month. The falling snow blurred the blue and white Christmas lights twinkling in the maple and fir trees, and the light of her camera illuminated the piles of snow that had already accumulated that evening.

"Another 20 centimetres of heavy wet snow tomorrow here in southern New Brunswick, close to 30 centimetres farther north toward Perth-Andover and Grand Falls." Shelley shivered as she delivered the last item in the forecast. The wind was cold and out of the northeast. "Another weather system will bring more snow early next week. Keep your shovels handy and get some more gas for your snowblower," she said, signing off from Moncton.

The red light on the camera in front of me came on.

"Thank you, Shelley," I said.

I was standing beside a large flat-screen television that displayed a map of power outages in the Maritimes. Behind me you could see the rest of our newsroom and the team that produced the weekend morning show.

"It's another snowy weekend in Atlantic Canada," I said. "This is the third snowstorm in the past ten days, and more snow is coming to start the week on Monday. You can't catch a break this winter. This brand of heavy, wet snow and the strong wind have led to power failures in the Fredericton and Woodstock areas. NB Power reports nearly five thousand customers are in the dark this morning." The map changed to reflect highway conditions. "Roads are snow covered, slushy, and not expected to improve until later this afternoon. Take it easy if you absolutely must travel today."

My producer, Rania Walker, counted me into the commercial break as the camera light blinked off.

La Niña had developed the previous September and was now, six months later, at its peak. La Niña is a cooling of the surface water in the Pacific Ocean along the coast of South America. It is a weather pattern that occurs on an increasingly frequent cycle and impacts global weather patterns.

In Atlantic Canada, a La Niña year usually produces slightly milder winters with average amounts of snow. This year the forecasting was nearly bang on for milder weather. However, the track of the jet stream was positioning more storms than expected into eastern Canada. The result was often heavy wet snow that mixed with rain in New Brunswick.

The reports from New Brunswick became a fixture on The Weather Network in February and March 2008. Shelley joined us from the base of Crabbe Mountain near Fredericton, where skiers and snowboarders were slaloming down neatly groomed trails and enjoying fresh snow over the March Break. Shelley reported from the back of a Ski-Doo as snow-mobilers rode the trails and revelled in the winter wonder-land. Most of New Brunswick receives more than 2 metres of snow every winter; in the northern part of the province, nearly twice that. This winter, some areas had almost 4 metres still on the ground in mid-March.

In fact, farther north, in the wilderness that stretches from northern Maine to La Baie-des-Chaleurs and the Témiscouata region in Quebec, the snowpack was nearly 50 percent above average. The snow was deep enough that some sugar bush op-erators had to dig out the bases of their maple trees to begin preparations for the spring tapping for maple syrup.

In Fredericton, provincial government officials were grow-ing concerned about flooding in the coming months. The long-term forecast was pointing to a sudden warming in April and a high likelihood of above-average rain that month. It was a worst-case scenario for flooding in the Saint John River Valley.

When the Laurentide Ice Sheet retreated from the Atlantic coast thirteen thousand years ago, it left behind glacial drain-age routes, including the Saint John River. At 673 kilometres, the Saint John is the longest river in Atlantic Canada, and one of the longest along the Atlantic coast of North America. It drains an area of 55,000 square kilometres, and its sources are a series of lakes in northern Maine and eastern Quebec. For all its recorded history, the river has been known to flood with regularity in the spring.

The length of the Saint John River is divided into three sections. The upper basin runs from the headwaters to Edmund-ston, where the Madawaska River joins its flow southward toward the Aroostook River close to the town of Perth-Andover. The middle basin flows past Fredericton toward Gagetown and features wider valleys of hardwood trees and rolling hills filled with rare vegetation, like wild ginger and wild coffee. Here the soil is rich and heavily farmed.

Fredericton, the capital of New Brunswick, has a metro population of over 100,000 people. Humans began to populate this area around 10,000 years ago, after the last ice age. When Europeans arrived in the mid-1500s, they met the Maliseet (Wəlastəkwewiyik) people, who had lived in the river valley for thousands of years. The Maliseet were more agrarian than their Mi'kmaq brethren; they knew the habits of the river, cultivating its shoreline and lowlands while living in the hills above the floodplain.

When European colonists began settling in the river valley in the 1600s, they built their farms along the river in the French seigneurial system of land allotment. Long, narrow tracts extended inland from the river, similar to the way land was partitioned in Quebec.

The Acadians who settled the area quickly learned that flooding would preclude living near the river. The first written accounts of flooding are from 1696 in Jemseg. The settlement moved to higher ground when the river rose to cover the fields and community that spring.

The French colony was called Fort Nashwaak and later Pointe Sainte-Anne. It was built on the north side of the Saint John River, across from present-day Fredericton, and would become the capital of Acadia. More French colonists arrived, and the settlement grew until the beginning of the Seven Years' War in Europe in 1756.

In North America, that war was fought as the French and

Indian War and pitted the French against the British, both of which recruited Indigenous Peoples to fight on each side. By the end of the war in 1763, the British had captured much of the French territory in North America, and the resulting Treaty of Paris divided the colonies of North America and paved the road toward Canada as we know it and the independence of the United States.

Following the war, the French Acadians were expelled from the Maritimes, leaving the land for British immigrants to settle. The American Revolution was also taking place at this time. That war delivered an onslaught of United Empire Loyalists—Britons loyal to the Crown and opposed to independence in America—to New Brunswick.

The Loyalists found a haven in the lush farmland around the old Acadian settlement of Pointe Sainte-Anne, which they renamed Fredericstown after Frederick, son of King George II of England. The town became the capital of the new colony and quickly expanded; streets were laid out in the typical British grid pattern in the lowlands along the mighty Saint John River.

Growing up in the Maritimes, I visited Fredericton on many occasions, particularly for football games between the UNB Red Bombers and my Acadia Axemen. King's College, now the University of New Brunswick, was founded here in 1785, making it the oldest English-speaking university in Canada and the eighth oldest in North America. It is the cultural and governmental heart of the province. The downtown area is made up of many old buildings that date back to the early 1800s, and the entire lower town is built on the floodplain.

Walking south, past the old garrison buildings on Carleton

Street and then away from the river on Regent Street, the land is flat and doesn't begin to rise until you reach Beaverbrook Street, more than a kilometre from the river. Then the roads climb uphill and out of the floodplain. Homes on the tree-lined hillside streets date back to the beginning of the twentieth century.

When you reach the crest of the hill at Priestman Street, you find Uptown, or the Fredericton of the last eighty years. Strip malls, auto dealerships, chain hotels, and fast-food outlets line the broad four-lane boulevard. This is what most travellers see when they pass through. The beauty of this city and much of its population are at the base of the hills, along the river.

The Saint John River has always flooded. Like in Manitoba, our human footprint grew into a vast floodplain as we converted more land for agricultural use and expanded our towns and cities. Our incursion has led to costly problems.

The final 120 kilometres of the river is considered tidal from Fredericton to its mouth at Saint John, at the end of the lower basin. Here the river ends its 673-kilometre journey at the famous Reversing Falls and empties into the Bay of Fundy, home of the world's highest tides.

When the tide is out, the Saint John River rushes down a gorge, carrying 1,000 cubic metres of water per second over a series of rapids as it empties into the Bay of Fundy. When the tide is in, the gorge fills with seawater and the river reverses flow, allowing salt water to mix with the river until the tide changes and begins to fall again. Twice a day, every day.

The tide occurs for two reasons. The first is the rotation of Earth on its axis, and the second is the gravitational pull exerted on our planet by the moon. These two events create a bulge in the oceans' waters, causing them to rise and fall in a phenomenon known as "high tide" and "low tide."

The water rises when its position on Earth directly faces the moon. This rise in sea level, the high tide, is principally due to the moon's gravitational pull. The second bulge happens at the same time on the opposite side of Earth. There the mass of the moon and the motion of Earth's rotations create a centrifugal force that causes the water to rise on that side as well. The low tide is situated between the two areas of high tide, which occur once every twelve hours.

The extraordinary motion of the tide in the Bay of Fundy is also influenced by the physical shape of the bay and its seafloor. The bay, an inlet of the vast Atlantic Ocean, tapers and rapidly becomes shallow, which is what affords us the highest tides in the world. The water rises and falls 15 metres twice a day.

The massive changes and the topography at Saint John also force the water flow of the Lower Saint John River to change direction with each high tide. The river literally flows back upstream. During the flood season, this twice-daily reversal of the river's natural outflow to the Bay of Fundy exacerbates flooding and prolongs the time it takes for excess water to drain from the valley as far upstream as Fredericton.

Extensive floods occurred in 1923, when the river rose 8 metres above the winter low level in Fredericton. In the province's capital, flooding normally occurs when the river rises above 6 metres. Premier Walter Foster estimated the damage to be over $5 million. Fifty-seven bridges were damaged or destroyed, and the cost to repair and replace them was nearly $400,000, a huge sum at the time.

In 1934 the river would flood again, cutting off all roads

south of Fredericton to Jemseg. Two years later, in 1936, the river rose to 8.9 metres, flooding most of the downtown business area as well as the legislative buildings. It was the highest the Saint John River had risen in thirty-seven years.

The river continued to flood over the decades. Then in 1973 the big flood came. New records were set for flood levels as the river overflowed its banks in the upper basin at Edmundston and then Grand Falls. Perth-Andover and Woodstock were next as the swollen river expanded into the middle basin. In Fredericton the Saint John River rose to nearly 9 metres, overwhelming the downtown and causing heavy damage to the historic Lord Beaverbrook Hotel, as well as destroying an entire mobile home park. The water remained above flood stage for twelve days. The cost was almost $12 million; that number would be over $100 million when adjusted to today's market. The larger our urban footprint grew, the costlier the floods became.

A rapid warming in late April, combined with a heavier-than-average snowpack and two rainstorms over three days that produced nearly 100 millimetres of rain, led to the 1973 catastrophe. Now, thirty-five years later, in 2008, the climate and weather conditions were nearly identical, and Fredericton had grown to be an even larger city.

On April 1 the sun beamed brightly, and a steady southerly breeze filled the Saint John River Valley. There was 47 centimetres of snow on the ground in Fredericton as the temperature climbed to 7 degrees Celcius that afternoon. The next day was even warmer. Children were trading toboggans for bicycles and parkas for spring jackets. The snowpack was reduced by 7 centimetres in one day.

Water flowed as snowbanks shrank. The thermometer would

rise above freezing every day that week and then fall just below zero at night—perfect conditions to melt the winter snow in New Brunswick, Maine, and eastern Quebec.

The deep snowpack that covered northern New Brunswick was built up by snow that had fallen with a high water content. Some snow is fluffy and light; the flakes are small, sometimes almost like Styrofoam. Those flakes form in very cold air, and it's nearly impossible to make a good snowball with that kind of snow. Most of the snow in Atlantic Canada that winter was the heavy, wet, moisture-laden kind that is formed in milder air with more temperature variations. The flakes are large and often grow as they fall through air that is close to or even above the freezing mark. It's also the best snow for making a snowman.

This snowpack compressed as more snow piled on top of it, creating a layer of ice along the surface of the ground. The ice and frozen ground at the base would prevent melting snow from being absorbed in the earth during the thaw. Instead, it formed large pools of water that ran into low-lying areas, creeks, streams, and eventually rivers. In northern Maine, eastern Quebec, and northern New Brunswick, all that runoff was destined for the Saint John River.

By mid-April, the river was rapidly rising. The forecast rain had held off, but the bright sun and temperatures that reached above 10 degrees Celsius each day were melting the snowpack at an unprecedented rate. In Fredericton, all the snow on the ground had melted by mid-April. The Saint John River rose above its banks along the Riverfront Trail on April 22; the water was 6.5 metres above its usual level.

Upstream, ice jams were acting as dams, holding back large volumes of runoff meltwater. Unable to freely flow downstream, the water began rising. Already the area around City Hall in Edmundston had been flooded. The Madawaska islands were disappearing as the Saint John River swelled to flood the communities of Saint-Basile and Saint-Jacques; a day later Sainte-Anne-de-Madawaska and Saint-Léonard flooded.

The scale of flooding in the upper basin was an ominous sign for the inhabitants downstream. It would be only a matter of time until the ice jams broke up and released a surge of water into the already engorged middle and lower basins. Sandbagging had been underway for weeks in the Fredericton area. Now vulnerable homes along the river were being evacuated.

On Saturday, April 26, Shelley joined our weekend broadcast from Queen Street in Fredericton. The sky was blue as sunlight danced on the Saint John River behind her; the thermometer read 13 degrees Celsius. Shelley stood beside the majestic Beaverbrook Art Gallery. The river had submerged the walking trail and the back garden that led to the gallery. The streets downtown were inundated with floodwater and closed to traffic.

"There is real concern here that the river will rise beyond the peak seen in 1973. That would mean a new historic high water level, and already, residents up and down the valley have been alerted to prepare for evacuations at any moment," Shelley reported.

The maps that I displayed on television told the next part of the story. By Monday, April 28, three days of rain were in the forecast. A vigorous spring storm was moving along the east coast. There was an abundance of moisture in the atmosphere to feed the steady rain. It was the type of April storm that washes away the residual sand and dirt of winter and encourages the grass to turn green.

On schedule that same day, the rain arrived. For three days, heavy downpours fell across the region. The combination of increasingly milder temperatures and the additional rainwater flowing into the upper basin of the Saint John River began to weaken the ice jams that had been slowing the flow from farther upstream.

The anatomy of a river during the spring thaw is an interesting study. Each year, ice jams form in many of the same locations—where the river shallows, causing ice to become grounded on the riverbed; or where there is a narrowing or bend in the river, which leads to congestions of ice. These slight alterations to the river flow allow ice to collect, forming natural dams.

When the ice jams in the Saint John River finally broke, a massive surge of water was released into the middle basin of the river system. Instead of the river's usual summer flow rate of 1,000 cubic metres per second, now over 7,000 cubic metres of water per second were pouring downstream.

The onslaught forced NB Power to evacuate workers from the Grand Falls Generating Station as water quickly inundated the facility. Sixty kilometres downstream at the Beechwood Dam, all the floodgates were opened to allow the inbound volume of water to move through the sluices and to prevent the inundation that had happened at Grand Falls. Water raced through the Beechwood Generating Station and rapidly rose a metre. An hour later it had risen another metre and was still climbing. Five kilometres downstream in the village of Bath, the river rose along the steep embankments and flowed across the highway, forcing its closure. The village, with a population of five hundred, was cut off by the rising water.

Florenceville, headquarters of the McCain food empire, was next. Farms and homes dotting the highway beside the river were deluged as the flood surge continued its unstoppable course downstream. In a matter of hours downtown Bristol and Florenceville were flooded, a half-metre of water washing across the streets and then into homes and businesses.

On the drive south from Florenceville and Bristol on Route 130, the former route of the Trans-Canada Highway, the road traverses inland and away from the river as it climbs up the east side of the valley. Looking back up the valley, it was easy to see the scale of flooding. The river had breached its banks and was now forming an ever-growing lake over the wide valley. The flood was outlining the topography by filling all the low areas with dark, cold water.

Twenty kilometres south lies Hartland. The town was settled in 1797 by a New Yorker, William Orser, and its name is both a salute to its location in the agricultural heartland of the valley and in honour of James Hartley, who surveyed the area in the 1800s. Hartland is best known as the site of the longest covered bridge in the world. The nearly half-kilometre-long wooden span opened in 1901 and today is a provincial historic site.

The bridge was conceived in 1898 to connect the two sides of the river and eliminate the ferry crossing. The projected cost for an eight-span steel bridge was over $70,000, a price the community could ill afford. Instead, a seven-span wood bridge supported by wood piers was built at a cost of $27,000. Officially, the bridge opened in July 1901. However, it was first used by Dr. Estey on May 13, 1901, for a medical emergency across the river. Local lore says that holding your breath while crossing the bridge brings good luck and that in the days of horse and buggy travel, couples would often stop for a kiss mid-crossing.

The Hartland Bridge was nearly destroyed during the 1920 flood. Ice jams and high water caused two spans to collapse and fall into the river. Repairs lasted two years and saw the installation of five tapered concrete piers to replace the wooden ones. The redesign of the piers would forever help in disrupting the formation of ice jams at the bridge. Also added was the now famous wooden roof and walls. The bridge remained in constant use until the Hugh John Flemming Bridge was erected in 1960, just a few kilometres upstream.

On April 29, the river rose to less than 10 centimetres from the wooden beams that form the base of the bridge. Chunks of ice bumped against the cedar and pine roadbed. Historians and Hartlanders were gravely concerned that the bridge might not survive the torrent. The community watched from the banks of the Saint John River as their 107-year-old wooden bridge, closed to traffic, withstood yet another flood.

The surge continued its movement through the middle basin of the Saint John River Valley, raging in torrents thirty-six times greater than the usual summer flow, through the spillways at the Mactaquac Dam and its generating station. The water submerged riverside farms as well as much of Sugar Island and the chain of islands that lie in the Saint John River on the northern outskirts of the capital. The highest water was arriving at the lower basin of the Saint John River Valley. Here, the daily motion of the world's highest tides would have their impact on the spring flood.

Just after 8:30 a.m. on Thursday, May 1, the river crested at 8.36 metres above average in Fredericton. In the days leading to the crest, water had risen nearly a half-metre each day, slowly engulfing the riverside parks and then washing across the streets of downtown, flooding homes and businesses. Water lapped at the steps of the Legislative Assembly Building, and more than fifty streets in the lower downtown were closed. Eventually, the power flickered and went out.

Access roads to the two bridges were under water, and Fredericton was cut off to cars and trucks. It would be another week before the Saint John River began to recede. In Fredericton the water would remain above flood stage until May 8, a total of seventeen days, the longest stretch of flooding the city had ever seen.

South of Fredericton is the Maugerville–Sheffield floodplain. This flat, broad part of the river valley sits only 3 or 4 metres above sea level. Route 105, on the northern bank of the river, is built on a metre-high levee that holds back the river during non-flood periods, but for more than two weeks this area was submerged under a metre of water. Hundreds of families were forced from their homes, and livestock were relocated from farms.

As the rising water breeched the highway levee, it flowed across the floodplain to the northeastern treeline more than a kilometre inland from the river. There, the water spilled into the brackish marshland that feeds the Portobello Creek, pushing it to overflow its banks on its course to the province's largest lake, Grand Lake. The torrent of water that flowed into Grand Lake flooded cottages and farms along its shoreline. The lake continued to rise with excess floodwater. The only drainage out of the lake was the now engorged Jemseg River,

20 kilometres downstream near the site of the first recorded flood, more than three centuries earlier.

At Jemseg, the Saint John River meanders through a broad plain. The long lakes that reach into the adjoining valleys are all extensions of the river, creating a maze of waterways around low, flat islands. This is the tidal estuary: in this stage of the river, seawater flows upstream from the ocean approximately 10 metres beneath the more buoyant freshwater.

In the lower basin, the Saint John River rose anywhere from a metre to nearly 2 metres above flood stage as the deluge from upstream filled wetlands and backed up estuaries, making places like Coles Island, Long Island, and Upper Musquash Island slip under water. The river flooded into the community of Grand Bay and pushed its water upstream on the Kennebecasis River outside of Saint John, submerging the lowlands at Darlings Island and around the town of Hampton.

It would take nearly two weeks for the water to recede in the lower valley.

As the water retreated, it revealed the damage it had wrought: washed-out roads, potholes, eroded culverts, and damaged power poles. Sewer systems, bridges, roads, and railbeds all required inspection before they could be declared safe.

As is the case in so many floods, water quality and security were imperilled, gasoline and diesel fuel leeched from underground storage into floodwaters, sewer and septic systems purged effluent into the water. Agricultural waste and runoff, as well as common household chemicals, all found their way into the floodwaters of the Saint John River. The insured cost of the flood was over $23 million, though much more of the damage was uninsured, and many homes and businesses were written off as total losses.

As our climate warms, the atmosphere is able to hold more moisture and temperature gradients will become greater. Winter snowstorms and spring rains in Atlantic Canada will both become more severe in the coming years. That, in turn, will lead to more frequent and impactful flooding of the Saint John River.

CHAPTER 5

The Rock

Newfoundland, 2012

My maternal grandmother is from Bay Roberts, New-foundland, which is on the north shore of Conception Bay. She was a Dawe. The earliest Dawes arrived from Ireland in 1595, nearly three centuries before Canadian Confederation, and as a result the surname is the twenty-third most popular in the province. Perhaps that's why I am drawn to the weather-ravaged island, whose largest city, St. John's, is closer to Dublin, Ireland, than it is to Windsor, Ontario.

My career brought me to Newfoundland and Labrador several times; the first time was to cover Hurricane Leslie in 2012.

Leslie developed as a tropical depression, a low-pressure system in the southern latitudes, early in the morning of August 30, 2012, some 2,000 kilometres west of the Lesser Antilles, which is in the middle of the Atlantic Ocean. By the end of the day Leslie would become the twelfth named storm of the year, with winds clocking in at over 100 kilometres per hour.

For several days in early September, the tropical storm moved slowly westward across the Atlantic. A strong ridge of stable high pressure over much of the Caribbean and along the east coast of the United States slowed Leslie's forward motion, causing an upwelling of seawater beneath the storm. An upwelling happens when the circulation in a very slow-moving storm stirs the ocean, drawing cold water from its depths to the surface. This colder water deprives the storm of the energy it had been gleaning from the warmth at the surface of the sea, thus weakening the storm's strength and velocity.

Leslie's strength oscillated during this upwelling phase; winds varied in velocity from just over 60 kilometres per hour to as high as 130 kilometres per hour. As the storm moved slowly northwest, past Bermuda, its wind field expanded to a radius of 1,800 kilometres, with tropical-storm-force winds extending nearly 300 kilometres from the eye.

By Sunday, September 2, Leslie was being drawn toward a large upper atmospheric depression positioned over the Labrador Sea. The storm had raced past Bermuda, causing minimal damage, and was expected to hit Newfoundland on Tuesday, September 4.

I was thousands of kilometres from Newfoundland, speaking at a risk management conference in Winnipeg.

"Who picks the names for hurricanes?" a middle-aged man in a blue golf shirt and khakis asked during the question-and-answer session. I had just given a talk about overland flooding during tropical storms and heavy rainfall events.

"That's one of my favourite questions," I replied. "We only began our current formal naming process in 1950."

During the Second World War, storms were given female names for the benefit of both the sailors and the airmen who crossed the Atlantic. Prior to that, forecasters would name a storm only if it had been especially impactful. In 1950, the National Hurricane Center (NHC) in Miami, Florida, began the tradition of naming hurricanes, in order to streamline messaging about severe weather. They chose short distinctive names that were easy to remember and that followed the old phonetic alphabet: Able, Baker, Charlie, and so on.

By 1953, the phonetic alphabet had been globally standardized to what we know now: Alpha, Bravo, Charlie, Delta, etc. But instead of following the new phonetic alphabet, the NHC decided to use an alphabetical list of female names. Alice was the first tropical storm using the new naming system, in late May of 1953, when she struck Cuba and Florida. To make the naming tradition less sexist, and to broaden the pool of names, the NHC began using a list of alternating male and female names in 1978. The first storm to be given a male name was Hurricane Bob, which made landfall on the U.S. Gulf Coast in 1979.

The World Meteorological Organization (WMO), based in Geneva, Switzerland, manages the names of all tropical storms, hurricanes, cyclones, and typhoons around the world. The naming of storms in the Atlantic Basin, Eastern and Western Pacific, and Indian Ocean, both north and south of the equator, is the responsibility of the lead weather agency in the area. For our region of the globe, the Atlantic Basin, the National Hurricane Center provides the names of major storms. In each region the names reflect the predominant local languages.

"The names that we use come from one of six lists that the WMO uses in rotation. The names are chosen alphabetically and rotate between male and female names, as well as English, Spanish, French," I said. "The list of names used in 2010 would be the exact same names we use again in 2016, unless the WMO decides to retire a name."

Storms that cause catastrophic damage and death are permanently retired by the WMO. Andrew, Katrina, and Juan have all been retired and replaced. When Andrew was withdrawn from use, Alex was the replacement; Katia was the name that replaced Katrina; and Joaquin took over for Juan.

"There are no names in all three of our languages that begin with the letters Q, U, X, Y, or Z," I continued, "so the list is always twenty-one names long. If there are more than twenty-one storms in a given year, then we begin using the Greek alphabet—Alpha, Beta, Gamma, Delta, Epsilon, Zeta, and the rest." I paused for a moment. "The first time that we had to use the Greek letters was in 2005, which was the most active year in history, with twenty-seven named storms. Forecasters used the first six Greek letters, up to Zeta. That was the year of Katrina, Rita, and Wilma, all so destructive that their names were retired."

I stopped and looked at the room of professionals who specialized in disaster mitigation. We all knew the gravity of increasingly powerful and frequent weather events, and were aware of the likelihood of Hurricane Leslie making landfall on the east coast in the coming days.

The Weather Network had been closely monitoring Leslie. Our concern was that this storm might be similar to Hurricane Igor, which had hit Newfoundland just two years earlier. Igor was the costliest hurricane in the province's history, the third wettest in Canadian history, and the strongest to strike Newfoundland since 1935. I had a conference call that evening with Peter Bozinov, our executive producer at The Weather Network. We both felt that this was an important story to cover and a good opportunity to test some newly purchased remote broadcast equipment. Cameraman Dwayne Oud and I would leave for St. John's first thing Monday morning.

The network enjoyed good television ratings, especially during a storm or an impending storm. Over the past decade, The Weather Network had been doing more live reporting from the field, but our ability to broadcast live was limited to where there was a Bell Telephone landline access point. These broadcast access points are in several locations in most large cities across the country and are utilized through a booking and fee arrangement. We used three locations in Toronto for live reports every morning on *Good Morning Toronto*. We could go live, but not everywhere.

In the spring of 2012, a solution arrived.

The Weather Network had just purchased a Dejero unit. This new technology enabled the use of mobile cellular telephone links to carry live television signals. The Dejero system allowed for the first live, ongoing transmission of an Olympic Torch Relay during the 2010 Winter Olympics in Vancouver. Many broadcasters had begun using this new system to augment their live broadcasts with satellite trucks. At The Weather Network, we would have the opportunity to test the limits and durability of this technology by using it in a hurricane.

Dwayne and I arrived in St. John's on Monday afternoon,

September 10, about eighteen hours ahead of the storm. We rented a Chevy Blazer at the airport, loaded it with our equipment, and made our way into downtown.

It was overcast but mild as we drove along Kings Bridge Road. We pulled the Blazer into the Colemans Market parking lot near Quidi Vidi Lake. We knew we would need extra batteries for the microphones and agreed we should buy some water and food in case the power went out during the storm.

"Well, look who's here," the cashier said as we piled our groceries onto the counter. "Here for the storm, are you? So you think it's going to be a bad one then?"

"Could be," I replied. "It's a hurricane. They can be tricky, you know, the wind and the rain, and it's been just a couple of years since Igor did all that damage."

Then she said what has since been repeated to me so many times, by so many Newfoundlanders: "You should have been here for the last one we had."

Make no mistake about it, these words are said for the benefit of those, like Dwayne and me, who "come from away," which means we aren't native to the island and, fairly, aren't inured to the local climate. And what the cashier meant is "This might be bad, but we've survived worse." Extreme weather is a part of life here, and surviving it is a badge of Newfoundland honour.

Newfoundland and Labrador are weather record makers. Gander is the snowiest city in Canada, with 4.5 metres of snowfall in an average winter. Gander and St. John's are tied for the most hours of freezing rain per year. St. John's is the foggiest city in Canada, with fog reported 206 days a year; it's also the windiest, with an average hourly wind speed of 21.9 kilometres per hour.

Dwayne and I headed up to Signal Hill National Historic Site. Here the precipice on the north side of the narrow entrance to St. John's Harbour rises nearly vertically from the ocean. At its crest are fortifications that go back to the mid-1700s, and Cabot Tower, which opened in 1900 and is close to where the first transatlantic wireless message was received in December 1901.

We set up the camera and Dejero unit beside Cabot Tower. I looked through the viewfinder and saw the sun setting on the horizon, bathing St. John's in a golden light.

"This'll look awesome," I said as Dwayne dialled up our Oakville control room.

The Storm Centre introduction played in my earpiece and Robyn Yakiwchuk, our producer, counted me down: "Two, one, and you are beside Suzanne in a two-box screen." Suzanne was on our Storm Centre set in one animated computer-generated box; I was in the other, with the caption "Live from St. John's." It was a great scene; the city looked wonderful in the background.

"I want to tell you a really cool story," I began. "This is Signal Hill. Right here, 113 years ago, Guglielmo Marconi and his assistant George Kemp flew a kite that had a 150-metre wire tail. They were hoping to hear the Morse code letter S that was being continuously transmitted from an antenna 3,400 kilometres away in Poldhu, in Cornwall, England," I continued as I walked around Cabot Tower. "Marconi heard the signal first at 12:30 p.m., and then fifteen more times that afternoon."

I stopped walking. Behind me the vast Atlantic Ocean sparkled in the setting sunlight. "Today we take for granted a phone call to or from anywhere in the world. This TV signal is being transmitted 2,100 kilometres through a modified cellphone to our Oakville studio, then via satellite to your local cable company, which sends it through a wire to your television all in one second. How cool is technology!"

We wrapped up the segment, packed up the truck, and drove to the harbour. We set up our camera and Dejero on the main pier that parallels Harbour Drive. A cruise ship was berthed, and about a hundred tourists were making their way along the promenade to the vessel.

"This is a great location to go live from," I said to Dwayne.

The shot looked nice on TV, with the imposing cruise ship as a backdrop, and the hustle and bustle of people boarding the liner on covered drawbridges. Suzanne iterated the latest information on the track and strength of Tropical Storm Leslie, and then I spoke with the tourists.

"We are a bit disappointed. We wanted to see more of this beautiful city," said Francine, a well-dressed woman in her fifties.

Her partner, another youthful-looking fifty-something woman, added, "We're leaving early. The purser told us that it's better to ride out a storm at sea, but I'd rather be here on dry land for it myself."

A large ship is often much better off at sea in a storm than moored in a harbour and facing potential grounding or dockside damage.

The wind began to pick up and intermittent rain fell as the sun was setting. Lights twinkled on the water. A group of conference-goers stopped to talk with us on the pier. They were on their way to get screeched in. Being screeched in is a tradition and rite of passage for tourists who visit Newfoundland and Labrador. The bars on George Street in downtown St. John's do a swift trade in the ceremony, which makes one an honorary Newfoundlander. The ritual involves donning a sou'wester hat, doing a few jigs with an ugly stick, kissing a codfish, and drinking a few shots of Newfoundland rum called "screech." There is much hooting, hollering, and encouragement throughout the hallowed rite.

It would have been fun to watch the group, in their casual office attire and with conference ID tags strung around their necks, as they followed all the strict protocol and procedure of the ritual. But we had to continue our live broadcast on the pier. Suzanne and I talked about when the worst weather would come. Behind me, the massive Holland America cruise ship slowly steered out of the harbour.

"They probably won't be serving soup tonight," I said to Suzanne as the liner's stern disappeared through the narrows and sailed onto the open Atlantic Ocean.

By late that night, Leslie was due east of Cape Breton, Nova Scotia, and was still a tropical storm with winds that were reaching 100 kilometres per hour. It had just passed over much cooler ocean waters and was beginning to interact with a strong burst of colder air from eastern Canada. Leslie was becoming what is referred to as an "extratropical cyclone." These are large-scale low-pressure systems that often extend more than a thousand kilometres outward from their centre. Leslie was following a natural cycle for hurricanes as they move to more northern latitudes—its tropical characteristics were being replaced by those of a strong mid-latitude storm, with very heavy rain and powerful wind.

Unlike a hurricane, which gathers energy from warm ocean water and well-aligned atmospheric winds, northern storms gain strength from a clashing of cold and warm air. The wind can be as strong as many hurricanes, and the rain or snow equal in volume and velocity to the precipitation of a tropical storm. The difference is where the storm forms.

Dwayne and I wrapped up our live broadcast at ten in the evening and headed back to the hotel. Rain was falling steadily; the wind was warm and strong, coming from the southeast at 50 kilometres per hour. On our drive through the narrow, twisting streets, we agreed that the best location to find the strongest wind would be at Cape Spear, the most easterly point in North America.

At 3 a.m., I woke to the sound of rain lashing against my window and the trees creaking and straining against the wind. The red digital readout on the alarm clock told me the power was still on. I checked my laptop for the latest weather observations from the St. John's airport; the wind was steady at 60 kilometres per hour with gusts at 75 kilometres per hour. The satellite image displayed the sprawling storm with a circulation centre about 200 kilometres south of the Burin Peninsula.

Dwayne and I met at 4:30 a.m., loaded our equipment into the Blazer, and drove through downtown St. John's on Water Street. The wipers kept time with the rain while the wind buffeted the truck as we passed shops and pubs and then past the container pier and drydocks. The GPS directed us onto Blackhead Road, and we drove east toward the faintly lightening sky.

Newfoundland can be best described as vast and beautifully rugged with sporadic human habitation. The city quickly vanished in our rear-view mirror. We followed the road as it wound through treed hollows and over barren, windswept hills. There were few indications that we were less than 10 kilometres from the provincial capital.

We drove the truck up a steep hill to the parking lot at the Cape Spear lighthouse. We were as far east as you can go in Canada. Here, the land is exposed to the elements.

The early Portuguese fishermen and seafarers began coming to this area shortly after John Cabot arrived in 1497. They named it Cabo Da Boa Esperança—Cape of Good Hope. Over time, it was called Cap d'Espoir by the French and later became Cape Spear to the English-speaking settlers.

The original lighthouse was completed in 1836. It was the second one built on the island; the first was at Fort Amherst, at the entrance to St. John's Harbour. The new light at Cape Spear was on top of a three-storey white building; the beacon had come from Inchkeith, Scotland, and was lit by oil. A foghorn was added to the site in 1878 to help guide ships toward St. John's Harbour. The new concrete lighthouse was built in 1955 and stands not far from the original.

Rain and wind shook the truck. Across the parking lot we could see a pickup truck rocking in the wind. Its occupants were eating doughnuts, drinking coffee, and watching the massive waves heave and roll toward the shore before crashing against the rocks some 75 metres below.

When I opened my door, the wind wrenched it from my hands.

"Holy shit," I said.

"Can you close it?" Dwayne replied. Papers were blowing out the door and flew through the air.

As I forced the door shut, I was peppered with blowing sand, little bits of gravel, and high-velocity rain. Dwayne turned the Blazer so that it faced the wind. I opened the rear door and got back in the truck. Those few minutes outside had nearly soaked my jeans.

"We're going to get drenched here," I said.

"Great," Dwayne replied. He called the office to let them know we'd be ready to go live in ten minutes.

The driving rain and wind, as well as the remote location, would be a good test for the Dejero's capabilities. Dwayne and I fought against the wind as we lugged our gear up the long pathway to the original lighthouse.

We crouched down beside the lighthouse to evade the wind and rain, connected the camera to the Dejero, and dialled up the control room in Oakville.

"Hi, we're all set. It's windy here," I said to Jamie Hall, one of the producers at TWN.

"Good, the signal is strong. The Dejero is working nicely," he said into my earpiece. "You're on with Chris Murphy in thirty seconds."

The wind roared as I stood on the highest ground I could find; the open ocean with its massive rolling swells was behind me. Keeping my balance was a struggle as powerful gusts blew upward along the cliff face. I could taste the saltiness of the sea spray mixed in with the rain.

The audio feed was difficult to hear over the loud, rumbling wind. I had to turn up the volume to make out Chris's words and Jamie's directions.

"Good morning from the easternmost point in Canada, and we are less than 100 kilometres from the centre of this storm Leslie," I said. "Because we are on the northeast side of the storm, the strongest winds will occur over the next three hours. Right now, Leslie is racing to the north and quickly transforming from a tropical storm to a nasty nor'easter."

We discussed the potential for damage but underscored that a storm of this strength was quite common in Newfoundland. Cool, dry air from the continent meets moist marine air here, and two ocean currents—the warm Gulf Stream and the cold Labrador Current—meet in the waters that surround Newfoundland.

This combination of contrasting temperatures in both the ocean and the atmosphere needs only slight trigger mechanisms, like a cold front, to rapidly incubate and deepen low pressure. It's why you so often hear of "weather bombs" when there is talk of a storm in Atlantic Canada. A weather bomb occurs when the barometric pressure at the centre of a storm falls 24 millibars in 24 hours. That steep, rapid decline is indicative of accelerated wind and precipitation, usually fuelled by an infusion of cold air.

Dwayne and I decided to move closer to the cliffs for our next live report. We wanted to show both the beauty of the landscape and the size of the waves as they crashed against the base of the promontory. The constant screaming of the wind had become a monotone, and the rain had been pelting us long enough that we were both soaking wet despite all our rain gear.

The camera was on Dwayne's shoulder. He stood with his knees slightly bent and legs apart for good balance against the wind.

In our earpieces Jamie counted us into the live segment: "Three, two, one, on you."

I held an anemometer above my head and said, "This instrument measures wind speed. It's showing 99 kilometres per hour right now."

While I was speaking, a man in a black rubber scuba suit, flippers, diving mask, and snorkel walked past Dwayne and stood next to me.

"Hey b'y, where ya longs to," he said. He was holding a microphone too. "I'm JLaC from K-Rock. Whatta y'at? Is ya'ard at it all de time or wa?"

We were both doing live reports at the same time. It turned out to be good fun on both The Weather Network and the K-Rock morning show in St. John's.

JLaC explained that he'd seen us on TV and was coming out this way for his morning swim. He wondered what all the fuss was about with the weather.

"G'wan, ya shoulda been here fer da last one we had," he said.

If it looked like a sketch that Mark Critch would do on *This Hour Has 22 Minutes*, then that's because the host of the K-Rock morning show was Mike Critch Campbell, the equally funny brother of Mark. It had been his idea to join us.

At 8 a.m., after broadcasting from Cape Spear, Dwayne and I packed our equipment and drove back downtown. A half-dozen power poles had already fallen along the side of the road, and several homes in the Shea Heights neighbourhood were missing strips of vinyl siding.

We drove up Duckworth Street and parked in front of the

war memorial. From the park there was a good view of the harbour, and the street was lined with tall trees that had begun shedding branches. We exited the truck and began setting up the camera and Dejero for our next report.

Leslie had been considered a tropical cyclone when it had made landfall on the southern coast of the Burin Peninsula an hour earlier. Now a cold front was racing across the province, and it would redirect the remnant energy from Leslie into a growing area of low pressure located over the Labrador Sea. Leslie was no longer a tropical storm, and it would take several more hours for the wind to wane.

At 9 a.m., downtown St. John's was quiet and nearly empty of workers. As I walked along the middle of Duckworth Street, the wind blew shingles from houses and tarpaper from the roof of the brown brick Sir Humphrey Gilbert Building, a large multi-storey government office tower. A large sheet of tar-covered insulation blew overhead as I approached a century-old tree that had fallen onto a car.

"This is the type of damage St. John's has seen this morning," I said. "There are forty-five thousand customers without electricity, and several homes that were under construction in the Pleasantville neighbourhood have been severely damaged."

We next drove through the Bannerman Park and Georgestown neighbourhoods. Large trees blocked roads and had damaged some homes, but already people were out to clear the mess. City works crews with chainsaws, woodchippers, and dump trucks were quickly removing debris and reconnecting downed power lines.

I interviewed a couple who were clearing branches from

their front lawn. Their white picket fence had been smashed by the fallen limb of a great old oak tree. Their tools were simple, a rake and a handsaw. An elderly neighbour was sweeping the sidewalk as we spoke.

"It'll take some time to clean this up, but the city will be by soon to help out," Darlene told me. "You can count on them anytime there's a wind like this."

"It's just good people here," her husband added. "We look after one another."

Dwayne and I spent the next few hours talking with people who lived near Quidi Vidi Lake in the east end of St. John's. They were happy to come on TV and talk about the weather in Newfoundland and the storms they'd experienced in the past.

Rose told me, "If you don't like the weather out the front window, then look out the back door instead. And if that's not to your liking, then wait for five minutes and then check again."

We laughed at the truth in those words. The offshore currents, the surrounding ocean, and the topography provide Newfoundland with some of the most quickly changing weather in the country.

"Now, would you like to come in for a cup of tea and some toutons?" Rose asked us. The toutons, deep-fried bread dough served with molasses, were delicious, and Rose was an excellent host, sharing dozens of stories about her city with us.

By three o'clock the sun was shining, the wind had calmed, and it was a pleasant afternoon. It was September 11. Eleven years earlier, the people of this province had opened their homes and hearts to thousands of strangers who'd found themselves on this island on a terribly savage day. Those who had "come from away" learned of the warmth, generosity, and spirit of Newfoundlanders. Here, no matter what happens, people get up, dust themselves off, look for the positive, and use humour to help get on with living.

CHAPTER 6

Rush Hour

Toronto, 2013

Most very large urban centres have an impact on the behaviour of the atmosphere. The vast expanses of asphalt
and concrete that are our roads, highways, and parking lots, as
well as the countless dark roofs, gather and hold heat from the
sun every day.

On the hottest summer days, the temperature in the heart
of an urban area is often several degrees warmer than the surrounding countryside. At night, the constructed environment
retains the day's heat and will again record temperatures several degrees warmer than areas outside the city. In the winter,
the same principle applies, plus the buildings emit heat to
keep their occupants warm during the cold days and nights.

In the case of Toronto, the city's location also helps to
create a unique microclimate. The Greater Toronto Area,
or GTA, encompasses over 7,000 square kilometres of land
and is situated on the north shore of Lake Ontario, south of
Lake Huron and Georgian Bay. The land is relatively flat,
with most rivers and streams draining into Lake Ontario. The
western perimeter is defined by the Niagara Escarpment, a
200-metre ridge that extends from Niagara Falls northward

to the Bruce Peninsula. The lakes, the escarpment, and the massive urban sprawl influence the behaviour of the weather. And the greatest problems arise when there is heavy rain and thunderstorms.

In 1954, Hurricane Hazel poured 121 millimetres of rain beginning at noon on Friday, October 15. By six the next morning, the total was 225 millimetres. Only 10 percent of the rain was absorbed into the ground; the rest flowed into Toronto's streams, creeks, and rivers. Etobicoke Creek and the Humber River rapidly filled to overflowing, sweeping away bridges and roads. The flooding from that one event took the lives of eighty-one people.

As the city expanded and grew, more and more agricultural land and wetlands were paved over, forever changing natural drainage patterns and the courses of waterways. Sewer systems and spillways now direct runoff into several of the larger rivers in the area—the Don River, the Humber River, Etobicoke Creek, and the Credit River empty overflow water into Lake Ontario.

Monday, July 8, 2013, was again hot and humid, as it had been all weekend. At 8 a.m. the temperature was already 23 degrees Celsius, and the humidity made it feel like 30. Morning commuters crept south on the Don Valley Parkway; the main north–south artery descends 100 metres into the Don River Valley on its way into downtown Toronto. The trees that line the valley were lush and green. May and June had been wet. In fact, sixty of the previous hundred days had been rainy, and the ground was saturated with moisture.

A crowded GO train sped alongside the river, passing the clogged traffic on its way to Union Station. The Don River was flowing well above its usual July level.

Early in the afternoon, the southerly flow of tropical air had clouded the sky. Workers exiting downtown office towers walked into a wall of heat and humidity. The humidex indicated that it felt like it was 37.

Humidex is not an actual temperature measured by a thermometer; instead, it is a calculation of how hot the air feels when the humidity is taken into account. We stay cool by perspiring; the evaporation of our sweat helps to keep our bodies comfortable. However, high relative humidity means that the air around us is already heavily saturated with moisture and cannot absorb more. The evaporation of our perspiration is inhibited, leaving an unevaporated layer of sweat on our skin. That is what makes us feel uncomfortably warm on humid days.

Urban heat, rising from the city, was helping to lift the warm and humid air aloft. A gentle onshore lake breeze was flowing inland to replace the rising warm air mass. The same process was happening 150 kilometres to the north on the shores on Georgian Bay and Lake Huron. Warm air was rising over the land as the cooler air above the large lakes flowed onshore. A few kilometres inland from both lakes, the gently rising terrain further enhanced the push of warm air aloft.

By 1 p.m., large cumulus clouds were forming to the northwest of the city. These clouds grow to towering heights as warm moist air is driven upward by breezes, the water vapour condensing as it rises into cooler air a kilometer or so above the surface. As the clouds grow higher they become ominously darker, an indication of the great volumes of moisture contained within each cloud.

Conditions were now ideal for thunderstorms. A special

weather statement was issued, highlighting the unpredictability of the air mass and the threat of torrential downpours by late afternoon.

At 2 p.m., the first rumbles of thunder were heard in the suburban towns of Orangeville and Newmarket. Heavy rain fell and gradually drifted toward Lake Ontario. Severe thunderstorm warnings were issued to the public. Over the next hour, a line of thundershowers began moving southeast toward the city.

Around 4:40 p.m. a storm passed over the eighteen lanes of Highway 401 near Pearson International Airport, slowing traffic to a crawl. Rainwater began accumulating and flowing toward drains adjacent to the shoulder lanes. At times the water was deep enough to stall cars, and those that had pulled onto the shoulder were quickly submerged.

Traffic on North America's busiest highway ground to a halt for several kilometres in each direction. Five hundred thousand vehicles travel the highway every day, and at that moment thousands of commuters were stranded.

It was 5 p.m. Rush hour. Downtown, at Union Station, thousands of people were waiting to board the westbound trains to Long Branch, Port Credit, Clarkson, Milton, and points westward. Thousands more were boarding the east- and northbound GO trains for the journey home to Oshawa, Ajax, Whitby, Richmond Hill, and the numerous other cities that spread out from Toronto. A single train can accommodate almost 2,000 passengers. On one platform, 1,400 people began climbing onto the double-decker cars of train 835 bound for Richmond Hill, 40 kilometres north of Toronto. The train was scheduled to roll out of Union Station in twenty minutes.

Outside the downtown core, near the provincial legislative buildings at Museum Station, the air was sticky with the pungent smell of the underground. Workers in summer office attire jostled in the humid air that engulfed the tunnels. Screeching metal wheels echoed off the tiled walls of the platform as an automated voice announced the arrival of the next train on the Yonge–University line.

Dozens of doors slid open to expel passengers and take in new ones. On the street above, the rain had just begun. For the first five minutes, big drops slowly darkened the sidewalk. Suddenly the sky opened and released a deluge that flooded the street and sidewalks. Water began to pour down the stairs onto the subway platform and tracks. Passengers battled the river up the stairs, but when they stepped outside, they were met by pouring rain and water that eddied above their ankles. Others slogged through the water and onto subway cars. As the train pulled away from the station, they were relieved to leave the chaotic scene on the platform behind them. They exchanged bewildered glances as the lights began flickering and then were extinguished. The train slowed to a stop, the sound of rushing water magnified in the black tunnel.

In front of Union Station, pedestrians dashed for cover. A group of soaking wet people were huddled in the lobby of the stately Royal York Hotel, watching as rainwater filled the streets to the curb. Water pressure in the sewers threw manhole covers into the air. Five kilometres west, on Lake Shore Boulevard at Exhibition Place, the highway had flooded, and cars, trucks, and buses, even a fire truck, were stranded in the rising water. The floodwater ran like river rapids from the elevated decks of the Gardiner Expressway, spilling in great waterfalls onto the streets below.

Rivers of rainwater streamed down the ramps that led into underground parking garages beneath the high rises of the financial district. It wasn't long before the rising water began short-circuiting underground transformers.

The power was failing in many parts of the city. In less than an hour, 300,000 Torontonians would be without electricity; in the neighbouring city of Mississauga, nearly 600,000 people would lose power. Traffic was in chaos. Signal lights were out, and the power outage halted the city's streetcars, subways, and transit buses.

Asphalt and concrete can't absorb water; these materials are designed to direct liquid into the drainage system. In just one hour, the storm had delivered more rain to the city than Toronto usually gets during the entire month of July.

All that water was draining into Lake Ontario, but not quickly enough. The Don River overflowed in the lowlands north of Bloor Street, and in less than half an hour water had submerged the parkland and spilled onto the Don Valley Parkway. Hundreds of vehicles were floundering in water. Traffic came to a standstill. Many began to panic and tried to turn their vehicles around, but there was nowhere to go.

At 5:40 p.m., those stranded drivers looked across the Don River. GO train 835 from Union Station, carrying 1,400 passengers, moved along the rail line, slowed as it passed the Evergreen Brick Works at Pottery Road and Bayview Avenue, and travelled another 200 or 300 metres before stopping completely and becoming stranded in fast-flowing floodwater.

Commuters peered through the rain-streaked windows. They could see the motorists stuck in their cars. Other drivers had abandoned their vehicles and were wading through the water, northbound on the flooded Don Valley Parkway, seeking higher ground farther up the road.

Varying winds had colluded to bring about the torrential downpour. The thunderstorms that had formed north of the vast urban area began drifting south. High above Toronto, cold air rushed toward the ground, creating a strong wind. That wind pushed southward over the city and encountered the strong breeze coming inland from Lake Ontario. At the same time, a warm wind was flowing into Toronto from the southwest and being forced aloft by the southbound gust front, which led to the birth of new convective storms over downtown and simultaneously reinvigorated the storms that were drifting in from the north. Within minutes, in a series of meteorological events that was almost impossible to predict, the storm system doubled in intensity.

Across the city, emergency services were pushed to their limits. For several hours, hundreds of passengers were rescued from subway cars that had stalled in darkened tunnels and were taken in small groups to the safety of station platforms.

Police took up positions at intersections, directing traffic until the power could be restored. The public was kept away from dangerous locations where they could become injured by fast-moving water or missing storm grates or electrocuted by live wires. It took a convoy of Toronto Police boats to shuttle the passengers from GO train 835 to safety. The evacuation began nearly three hours after the train had become stranded and was completed by midnight.

City crews worked overnight to clear away mud and debris from the streets, highways, and sidewalks. Toronto Transit spent days repairing damage and ensuring the subways were safe.

Over a five-hour period, the city had received more than

126 millimetres of rain, and over half of that had fallen in less than two hours. It was the greatest one-day rainfall event in Toronto history. Incredibly, there were no deaths or serious injuries, but there was over $850 million in damages.

This wasn't the first time Toronto had experienced floods, and it certainly wouldn't be the last. In fact, overland flooding continues to dog the urban area, despite massive infrastructure investments to manage extreme rainfall events. In our warming climate, Toronto is expected to see its annual rainfall rise from 790 millimetres to more than 850 millimetres by 2050. Much of that will come in the form of singular extreme rainfall events.

Over the past twenty years, I have spent many days standing in the rushing waters of the Don, Credit, and Humber rivers reporting on floods. Cold water eddied just above my knees as I explained that tomorrow's weather had come too quickly for the fifty- to seventy-year-old flood management system in this enormous urban area. It was built without the knowledge that once-in-a-century floods would occur with much greater frequency—now once every three to seven years.

CHAPTER 7

The Bridge

Prince Edward Island, 2014

Wednesday morning, March 26, 2014. The bedside clock read 3:15 a.m. I'd set the alarm for 3:30, but I always wake before it goes off.

I got up and made coffee, which, in motels, always seemed to taste like it was flavoured with strychnine. My producer had sent an email overnight: CNN would like us to record an interview with them at 5:00 a.m. The storm had blanketed the east coast to Cape Cod, and it was their lead story. This storm had developed as a result of the interaction between a deep pool of cold air that sank southeastward from Quebec and a surge of unusually mild air that had been drawn northward along the eastern seaboard of the United States. The low-pressure centre had begun to rapidly deepen as the system passed over Virginia. It was already snowing heavily in New York and Boston. CNN wanted a follow-on story tracking the storm's path into Canada, as conditions here were forecast to be even more severe.

I looked out the window. It was still dark and it hadn't started snowing.

My laptop was open and I checked my bookmarked

weather information sites. There was no snow here in Charlottetown. Yarmouth, Nova Scotia, had moderate snow and strong winds; Saint John, New Brunswick, was reporting light snow; Moncton, New Brunswick, 120 kilometres away, was overcast. A storm always looks better on television when it's actually stormy. Computer models for this storm still forecast snow in Prince Edward Island by midday.

New Brunswick, Nova Scotia, and Prince Edward Island—the Maritimes—are an hour ahead of the Eastern time zone. Newfoundland is another thirty minutes ahead. Six in the morning at our studios in Oakville is actually seven local time in Charlottetown, which meant there would be good natural light when we went live with our reports.

I called my cameraman Dwayne at 5 a.m. "Let's find a blizzard, buddy" were the first words I said when he answered. He was already up getting our equipment ready. His room was a sea of cables, camera batteries, chargers, two tripods, the bulky old-school Sony camera, and the amazing Dejero box that allowed us to be live on television without a satellite truck. When you're 1,300 kilometres from the office, you carry two of everything.

A day earlier we had flown into Moncton. As is usual for a snowstorm, we rented the biggest truck we could get. This time it was a Yukon. On the drive to Charlottetown, Dwayne and I formulated how we would shoot a couple of items on this impending storm. We decided that the Confederation Bridge would provide a great backdrop to the story, and its importance to Prince Edward Islanders would become the essence of my conversation with Suzanne on The Weather Network.

The "Fixed Link," or Confederation Bridge, was designed

and built specifically with winter storms in mind. In 1891, it was George Howlan, a shipping merchant and Island politician, who first proposed a fixed link: a railway tunnel that would connect the Island with the national railway system. Over the decades, proposals to link the Island to the mainland came and went. It wasn't until 1988 that a plebiscite was held, asking Islanders if they were in favour of a fixed link bridge. Nearly 60 percent of the province voted yes.

Construction began in October 1993. In 1997, the 12.9-kilometre-long box girder span replaced the busy ferry service between Cape Tormentine, New Brunswick, and Borden, PEI. The Confederation Bridge is the longest bridge in Canada, but it is also the longest bridge in the world that spans ice-covered water.

Before the bridge opened, life on the Island revolved around the ferries and the weather. A century ago, longboats with a crew of rowers would brave the icy strait to move goods and people brave enough to make the journey across the Northumberland Strait in the winter. The concrete span changed that. It was a permanent link that could be used year round in almost all weather. Restrictions on the bridge's use are imposed when the wind rises above 70 kilometres per hour. Wind speeds are constantly measured with a series of weather stations on the span. At a weather station, anemometers, which measure wind speed, are located several metres above the surface to avoid interference at ground level. The bridge rises to 60 metres above the strait.

When I was a child, my family took our August holiday at Tracadie Bay on the Island. It was a marvellous place to be in the summertime; the beaches, dunes, and tidal flats were a great playground for us kids.

One Saturday morning during Old Home Week, a celebration of the province's agricultural heritage, we met Stompin' Tom Connors and Bill Langstroth of *Singalong Jubilee* at the liquor store in Charlottetown. As always on the Island, CFCY radio was on in the background, with its ferry reports every half-hour.

I loved the ferry ride. During tourism season, there would be long waits, sometimes several hours, to get on board. So at the terminal there was a picnic area and restaurants with chips, deep-fried clams, and burgers. That was what life was like before the bridge was built.

With Confederation Bridge framed behind us and the wind picking up, we did our live shots with The Weather Network in Oakville. This inbound storm would be a weather bomb.

The strength of a storm is measured by several parameters. Atmospheric pressure is one. Think of the atmosphere as an aerial ocean with undulating waves. The crests and valleys in that ocean above us are high and low pressure. The centre of a storm is the centre of low pressure, which is exactly what it sounds like—where the weight that the atmosphere exerts is at its lowest.

Pressure is measured in millibars or inches of mercury. As the pressure lowers or falls, the measurement in a barometer falls. A drop in pressure usually occurs gradually. However, in the right conditions, it can drop rapidly. A drop of 24 millibars in 24 hours is called "bombogenesis," or a weather bomb.

Nature constantly seeks harmony, an equilibrium. When cold, dry arctic air surges into a moist, warm air mass, havoc prevails. Immense temperature variance on a grand scale creates a storm. The movement of all this atmosphere is what we feel as wind.

The sun was setting behind us as we drove across the bridge. The view from this marvel of engineering was spectacular. The Northumberland Strait was covered in ice.

The storm was due the next day, in the afternoon, according to the latest computer models, which had been accurate to within six hours on the expected life cycle and path of this storm. And so the next morning, Dwayne and I loaded the truck together. The swishing sound of our snowpants broke through the silence of the dark early hour. We were both wearing long johns, jeans, a snowsuit, two pairs of socks, warm boots, a corduroy shirt, a sweater, a parka, gloves, and a toque. It hadn't started snowing, and it wasn't even that cold yet. We looked ridiculous . . . for now.

We set up our gear on Grafton Street, right in front of the Confederation Centre of the Arts. For anyone who has been to Prince Edward Island, the building is instantly recognizable. Islanders stopped by to talk about the weather and what it was like before the bridge, when ice jammed the Northumberland Strait and only the icebreaking ferry MV *Abegweit* could transport food and other necessities from the mainland.

It started to snow at around 10 a.m. Big, soft flakes fell gently to the ground. In anticipation of the inbound weather, schools released students at lunchtime and cancelled the next day's classes. Businesses were looking at closing early. The last flights had flown in and out of Prince Edward Island. The province was shutting down and settling in to ride out the blizzard.

By definition, a blizzard is a sustained wind over 40 kilometres per hour for four hours or longer with snow reducing visibility to less than 400 metres. By three in the afternoon, snow was accumulating at a rate of 5 centimetres per hour, and the wind was blowing steadily from the northeast. It had gotten much colder. The temperature had been 0 degrees Celsius in the morning, and it had already dropped to –7. Visibility continued to drop, and there weren't many Islanders in downtown Charlottetown.

By nightfall, everything was closed or closing. The RCMP advised against any travel. The conditions were getting worse.

Since the heavy snow had begun falling, we'd been constantly moving our truck to prevent it from being snowed in. We'd find a plowed stretch of Kent Street and park there for an hour or so. Then as the snow built up we'd find another plowed section of Queen Street or Great George Street to set up our live shots.

We interviewed the local snowplow operators, who explained the methodology of road clearing. During a storm, the objective is to keep the main routes open for emergency vehicles. When the snow tapers off, the roads are usually cleared from the curb to the centre of the road. Large snowblowers then blow the plowed snow into dump trucks, which are unloaded outside of downtown. But this process doesn't work well when it's really windy. Like on this night.

At about nine that evening, we were recording a few of what we call "look lives," which would be used on The Weather Network late at night and overnight. Getting this bit of production done now would earn us a few hours of sleep before an early start the next day. We kept the information fairly generic, doing colour commentary on how the weather had behaved over the past day, and included information on the known school, business, and road closures.

The plan was to walk down the middle of a snowed-in Grafton Street in front of the well-lit Province House National Historic Site. Cars were drifted in to their rooftops. It was the perfect setting to illustrate the storm's majesty. The wind was blowing well over 60 kilometres per hour, and the snow was nearly blinding. Ahead of us, a blinking yellow light let us know that a lone plow was struggling with a snowbank.

We recorded one item and were set to do another when Dwayne couldn't get an image to appear on his camera. Only blue would show in the viewfinder.

"Weird. That's never happened before," he said.

We trudged back to the truck to call one of our technicians. After half an hour on the phone, we finally got the camera working again. Dwayne hoisted the Sony onto his shoulder, and I walked through the drifting snow to my spot down the street.

"Three, two, one . . . ," he shouted above the wind, and I began my report. Dwayne waved his hand at me. Through the blowing snow, he motioned me in his direction and I walked back to him. The viewfinder was blue again. We returned to the truck and made another call to our technician. We discovered that the snow had accumulated inside the lens and turned to ice, which likely froze the iris on the camera. The best possible solution, we were told, was to keep the camera in a warm, dry place for several hours, maybe use a blow-dryer on it if we had one.

But we still needed to deliver the "look live" reports. We had one completed and we needed three more. Dwayne then came up with what at the time was a brilliant idea. He would shoot the video on his new iPhone. This was before *everyone* shot and streamed videos from their phones on social media.

The next morning the camera seemed to be working, the coffee tasted the same, but the weather was different.

The wind peppered us with ice pellets and snow, the thermometer measured –17 degrees Celsius, and the clouds were scudding overhead with amazing speed. The storm had passed us overnight and was now northeast of the Island, and cold air was rushing across the Gulf of St. Lawrence behind it.

It took about twenty minutes to dig out the Yukon. We drove on fresh snow back to Province House to create the perfect before-and-after shot to illustrate the storm's effects. Restaurateurs spoke with us on camera as they shovelled the sidewalk. The plows were making fast work of the over 40 centimetres of snow.

Mid-morning, we left for the Confederation Bridge.

It was windy in the city, but the cover afforded by the buildings and trees had tamed the gusts and drifting snow. We drove out of Charlottetown on partially cleared roads. Twenty kilometres down the Trans-Canada Highway there was blue sky above us, but the world in front of us had vanished into a sea of white.

The strong wind was blowing snow, creating a fierce blizzard at ground level. We could see the tops of fences and trees, but everything beneath was simply white. Across the fields, smoke rose from the chimneys of farmhouses. We hadn't encountered any vehicles since we left the city.

We drove on slowly, but every 10 metres or so there would be a jolting thud as we hit another snowdrift. The wind was blowing some areas of the highway clear and amassing metre-high drifts in others. It took an hour and a half to drive the 28 kilometres to DeSable. We stopped there and waited to get back on the highway, behind a snowplow this time. Luck had come our way.

When we arrived, the bridge was still closed. The wind was clocking in at just shy of 100 kilometres per hour over the Northumberland Strait. Lights flashed red on the overhead signs. Some cars and transports had been abandoned in the lanes leading to the toll booths. On the transports, the mounds of snow rose to the tops of the trailers. The pavement was polished to ice. The wind was so powerful that it was a struggle to open the door of the Yukon.

We decided that the best way to shoot our live material was to park the truck with the front facing the wind. We opened the rear tail door and placed the camera inside the back of the truck to shelter it from the elements. The camera didn't freeze or become contaminated with ice, and we were able to capture the blowing snow and wind. The shot would become part of a promo that ran for several years on The Weather Network.

"I-it's so cold" were the first words I uttered as the wind relentlessly sprayed me with ice pellets at 90 kilometres per hour.

Beside Confederation Bridge is a service centre with a gas station, café, and museum. We packed up our gear and slowly drove the truck past the snowdrifts that lined the parking lot. There were about thirty cars and nearly as many transport trucks in the parking lot. The sun shone brightly, almost tauntingly, as it slowly sank in the sky.

The restaurant was busy, with people streaming in and out. We went inside to talk with the local folks. Their stories were fantastic. The majority had been waiting since the bridge had closed the previous evening. They were all stranded en route to Moncton to catch a morning flight to the Dominican

Republic; they were supposed to be in Punta Cana right now. The flight had been delayed and rescheduled for a late-afternoon departure. But the wind was still too strong to allow for safe passage across the bridge. All they could do was wait.

Maritimers, as a rule, take all things in stride and with a great sense of humour. Such was the case here with this group of sun seekers.

"I think this might be your fault, you know," a woman told me. She smiled and put her hand on my arm. "I'm just kidding, love."

We interviewed a couple from Stanhope Beach. They were newlyweds.

"My honeymoon was supposed to be fancy rum drinks, not this double-double from Timmies," the husband said to me. "But it's okay—we'll get there sometime, and we've got nice beaches here too. Just not right now."

By 9 p.m. the storm was moving rapidly away from Atlantic Canada. Its centre was about 200 kilometres northeast of St. John's. In its wake, high pressure was building, creating a more stable air mass. The havoc in the atmosphere ebbed. Later that evening, the wind speed gradually decreased and the bridge reopened, though too late for our sun seekers.

Dwayne and I climbed back into the Yukon and hit the Trans-Canada Highway for Charlottetown. The salt trucks had been out that afternoon, and bare pavement stretched ahead of us. As the sun set, a warm amber light glowed from the windows of the farms that dotted our drive. In the city the streets were slushy, and pedestrians navigated around snowbanks to cross the road.

"What a difference from last night," Dwayne said as we got

to the hotel. The parking lot had been plowed clean, and two men were shovelling away at a snowbank across the road.

The next morning the storm was becoming old news, and life on the Island was nearly back to normal. It was still cold, but the sun was up and the wind had eased. We packed our equipment, piled into the Yukon, and headed down the Trans-Canada, which was clear of snow. We drove up to the toll booth on the bridge. The transport trucks were gone, and only two cars were parked at the restaurant.

We chatted with the toll booth agent. She had worked the bridge since it had opened in 1997. In her recollection, this was only the second time it had been closed for longer than twelve hours. Previously, she had worked on the ferries. A storm like the one that had just passed would have shifted the ice on the Gulf of St. Lawrence and clogged the Strait, she said. It might have taken a few days to clear a passage route for the ferries.

Crossing the bridge takes about twelve minutes, lots of time to marvel at the expanse of ice on the Northumberland Strait. It's also enough time to reflect on the difference this bridge has made to life on the Island.

CHAPTER 8

The Beast

Fort McMurray, Alberta, 2016

My phone vibrated. I pulled it out of my pocket and saw that my wife, Susan, had just texted me.

"What's going on in Fort McMurray?" the message read.

One of her colleagues had taken a photo from the roof of Peter Pond Mall, the only shopping centre in the city. In the distance was a large plume of black smoke.

"Let me find out," I texted back.

It was Tuesday, May 3, 2016.

Since 2014, El Niño had been the dominating weather feature in the Northern Hemisphere. The pooling of warm water in the Pacific Ocean off the South American coast had had an impact on the weather pattern in North America, leading to drought conditions in the western United States. In Canada, it produced a milder and drier than usual winter in British Columbia and Alberta.

For much of Alberta, the past winter and spring had been decidedly unseasonal. In fact it was so mild that by April the Canadian Forest Fire Danger Rating had been elevated to high. There had already been 370 forest fires so far that year, double the provincial average.

The Fort McMurray fire had begun over the weekend, likely caused by a spark emitted by an all-terrain vehicle, and had spread quickly to the boreal forests that surround the city. Several neighbourhoods had been asked to evacuate. The fire was so far under control, but the weather was not promising.

My phone buzzed again with another photo from Susan, showing a black sky with massive columns of flame piercing through the darkness. The wind had shifted, and the fire was moving rapidly toward the city. A full evacuation had been ordered by Mayor Melissa Blake.

I sent the two photos to our executive producer Derek Snider, along with a text that read, "We need to be here ASAP."

For several decades, this part of Alberta had been trending toward mild, dry winters; winter snowfall was now nearly half of what it was fifty years earlier. The past winter had been the driest in more than fifty years with just 61 millimetres of melted snow and rain; the average is 132. Making matters worse, Fort McMurray had received only 1 millimetre of rain over the previous three weeks.

In addition, the average winter temperatures had grown to be about 5 degrees warmer than what they had been fifty years prior. The previous seven months had been the second warmest winter on record. The average temperature over the winter had been −4.5 degrees Celsius; normally it's closer to −8. The month of May, which was now just three days old, had seen the temperature climb to the high twenties and low thirties. The forecast projected a continuation of this record-breaking heat.

David Phillips, at Environment Canada, added these statistics to what is called the "spring dip," the time between the last snow melt and when the first leaves appear on trees. During this period the trees have their lowest moisture content, making them highly combustible. The fact that all of the snow had melted in Fort McMurray by March meant that the floor of the boreal forest was tinder dry. David also noted that the spruce, pine, and aspen in this part of Canada are a particularly volatile fuel for fire, which spreads easily through the crown, or top part, of the tree.

There was one other weather factor at play: an inversion.

The temperature usually decreases as you move higher up in the atmosphere, several degrees for every thousand metres. An inversion is when the opposite happens—the temperature in the atmosphere is warmer than it is at the surface.

Inversions are a common occurrence when high pressure is the dominant weather pattern. On a clear night, Earth's surface cools and warm air rises. The return of the sun in the morning hours reheats the surface, and by midday the inversion will begin to dissipate as the ground heats the air. But as the inversion breaks in the early afternoon hours, the recirculation of warm surface air causes winds to accelerate.

That's what happened in Fort McMurray.

I looked at the images on my phone again. The first displayed the smoke rising and then flattening out as it spread along the base of the inversion layer. In the second photo, the smoke and flames filled the frame. The inversion lid had dissipated or broken; the air temperature at the surface had become warmer than the air in the atmosphere. This hot air rose rapidly, creating strong wind gusts, and the newborn wind pushed the fire into unburned forest—new fuel for the burning flames.

The latest forecast charts showed no changes to the

weather pattern over western Canada for the next two weeks. The area of high pressure sitting over Alberta and Saskatchewan would remain in place; no rain was coming.

I went onto Twitter; #YMMFire was trending. People had posted videos of vehicles racing through the firestorm, some taken from the front passenger seat. Flames lapped at the windows, sparks and debris fell onto the hoods, and there was blinding black smoke everywhere. One video was taken from the roadside and captured lines of red taillights driving into a flaming, smoky abyss, embers falling from the sky and showering the ground. A wave of nausea rolled over me.

My phone vibrated with a message from Derek: "Pack for a week, we're booking flights right now." A second message appeared: "This will be the story of our time."

Less than twelve hours later, I was at Pearson Airport, waiting to board my flight to Edmonton. Passengers sat in unusual silence, looking up at the monitors. The news was filled with images from Fort McMurray. The entire city had been evacuated. Over eighty thousand people had fled their homes and communities.

Firefighters battled the flames on the streets and in the yards, doing their best to protect homes and businesses from destruction. Communities such as Beacon Hill, Timberlea, Blacksand Lodge, Waterways, Thickwood, Saprae Creek, and Abasand were under siege.

The fire was now being referred to as the Beast.

My colleagues Mark Robinson and George Kourounis were already on their way to Alberta and would meet our Calgary reporter Deb Matejicka for the five-hour drive to Fort McMurray.

I met up with my cameraman Shawn Legg outside the

terminal at Edmonton International Airport and we loaded up his Jeep with our gear. It was quite warm for early May; the air was still, and we had the windows down as we drove over the Queen Elizabeth II Highway toward 42 Street in Edmonton. Four hundred kilometres away, the RCMP had set up a barricade, closing access to Fort McMurray on Highway 63, the only route in and out of the city.

Mark, Deb, and George were already at the security blockade filing reports, and planned to sleep in their vehicle that night. I suggested that Shawn and I stay in Edmonton and report live in the morning from one of the many rescue centres that had been opened for the evacuees.

Shawn wheeled the Jeep into the driveway of the Stars Inn, across the highway from Edmonton Airport. The parking lot was filled with pickup trucks and cars packed with boxes, televisions, and oddly chosen appliances. Dozens of people were gathered in the parking lot, drinking beer and talking. We got out and walked through the crowd to check in. The place was packed; everyone smelled of smoke.

We began talking with a group that had come from Fort McMurray. It had taken them over twelve hours to find their way to the hotel. The traffic out of Fort McMurray had been gridlocked; so too the roads to Edmonton. Others had run out of fuel. Only the goodwill of strangers who had stopped to siphon gas from their cars had gotten them here.

At 4 a.m. the next day, I was standing in front of the A&W in Fort Saskatchewan, ready to go live on The Weather Network. The sun began tinting the eastern horizon with a hint of orange. Behind me, in neat rows, were more than one hundred jerry cans filled with gasoline. Handwritten signs and messages were propped up against the containers: God

Bless You, Free Gas, Thinking of You, Alberta Strong, YMM Strong.

Across the road, at the Tim Hortons drive-through window, a server was giving away coffee.

"It's for anyone from the fire and anyone who helps," he said. "It's the least I can do."

We drove to the Fort Saskatchewan Community Hall on 93 Avenue. Piles of clothing, shoes, food, diapers, and toys filled the halls and rooms. A dozen men and women were busy sorting the donations.

"We're doing whatever we can to help," one of the women told me. "They would do this for us. It's what we do. We're Canadians." The donations had come from people in town, she told us. Some they knew, others they didn't know; maybe some who were just passersby.

A young woman and little boy walked into the room; both were crying.

"Please can I just use your phone? I left mine at the fire." Tears rolled from the woman's eyes. "Please," she said again as I handed her my phone.

That afternoon, as we slowly made our way toward Fort McMurray, we pulled into the village of Boyle to fuel the truck. While we stood at the cashier, a man walked in. His dusty black Suburban was parked at the gas pumps with his wife, four kids, a dog, and seemingly everything of value stuffed into the back.

"I don't have any money," he said. "We left there so fast, I, we, didn't . . ." His voice trailed off. He looked down for a moment and then back at all of us waiting in line. He looked tired. He was wearing a day-old beard and a pained expression.

Behind me, a dump truck driver pulled on the chain of his wallet and quickly extracted three twenty-dollar bills.

"Here you go," he said. "God bless you."

We got back in the Jeep and drove along Highway 63. Just before Atmore, Highway 63 turns north and there you'll see a sign that reads, "Fort McMurray 240 kilometres." We followed the road until we arrived in Wandering River. There you will find a gas station, a motel, an A&W, a Burger King, and several cabins for rent. It is the last stop before Fort McMurray, which is about two hours farther along the highway. We knew that this was the closest accommodations we could get to the fire. The motel was sold out, but we were able to get a cabin with two bedrooms and a shared bathroom. It was rustic.

Done deal, we hit the road north again.

There is very little to see between Wandering River and Fort McMurray except for the highway, which has been dubbed the Highway of Death. Every year it claims victims during blizzards or in icy conditions. Too often the victim is a shift worker driving back to Edmonton or Calgary, tired after two weeks of hard work on the oil sands, or a transport driver, overworked and exhausted. The roadside is lined with trees—spruce, pine, alder, and low brush. The landscape betrays the scars of previous forest fires, and vast swaths of low-standing new growth show the regenerative power of the boreal forest.

On our drive, we saw abandoned cars, trucks, and camper trailers on both sides of the four-lane highway. All were heading in the same direction, away from Fort McMurray.

There was no other traffic on the highway, just us. We passed a motorcycle parked on its kickstand and a large travel trailer upside down in the centre median, its contents spilling from a ruptured wall. We hadn't seen a soul since Wandering River.

"Jesus Christ," we both said as the scene retreated in the rear-view mirror.

On the horizon, we could see the columns of smoke rising from the Beast. We were still more than 30 kilometres away, but our understanding of how large MWF-009 had grown in just four days became clearer. It had now consumed 85,000 hectares of forest.

The fire was burning close to the roadside. As we approached the RCMP barricade, we passed a line of several dozen vehicles parked on the shoulder—satellite trucks and what seemed to be the entire fleets from newsrooms across the province. Several RCMP cruisers, their lights flashing, and two dump trucks were blocking any further access along the highway.

We stopped in front of a police car.

An officer walked up to us and asked for our identification. We explained that we were part of the media here to cover the fire. He gestured to the line and stated very clearly and slowly, "Go back there, get in the line, do not go into the woods, stay away from the fire. We can't be responsible for your safety here; we will keep you informed on a regular basis, though."

We turned the Jeep around and drove back to the line of brightly painted media vehicles. Mark, Deb, and George were in the Weather Network Alberta truck, which was parked midway along the parade of rigs. We stopped beside them on the opposite shoulder, and together the five of us sat on the roadside observing the scene around us.

Behind us trees burned, some of them exploding in flames.

A 10-metre-tall spruce turned into a fireball and was gone in ten seconds. Less than 15 kilometres away, firefighters from Fort McMurray and scores of other western Canadian communities were trying to stop the flames from consuming more of the city. Some battles would be won, and others would be lost.

Far more dangers confronted the fire crews in the urban setting of Fort McMurray. Barbeque propane tanks exploded, sending shrapnel in all directions. Ammunition would go off, emitting the repeated sound of gunshots. Fuel in furnace tanks ignited; natural gas lines burned. Cars, trucks, and ATVs would burst into flame when the fuel in their tanks reached the flashpoint.

The volume of water used to combat the flames was exhausting the supply. Water from the Athabasca River was pumped directly into the water system, bypassing the treatment plant and contaminating the entire civic water system.

As dusk began to fall, we did a live report on The Weather Network, then another for the CBC.

"No," I said, "there will be no rain, not for a long time."

Behind us as darkness descended, the fires bathed the sky in an eerie orange. The nighttime inversion was building, and the wind had calmed again. There would be several good hours for the firefighters to work now. That night, entire streets of unburned homes were bulldozed to create fire breaks to keep the Beast from burning more of the townsite.

At four the next morning, we began our live broadcast from the gas station parking lot in Wandering River. The sun was slowly lighting the sky. The satellite trucks that had been at the barricade were parked in front of the Wandering River Motel. This was where most of the television crews were bunked in. It was warm and dry, and the wind would build during the day. Distant flames illuminated the early morning sky, and plumes of smoke were backlit by the rising sun.

The first car arrived at the gas station at around 8 a.m. Then dozens of cars, pickup trucks, and SUVs began pulling off the southbound lane of Highway 63 and into the parking lot. Moms, dads, kids, and pets emerged looking tired and dirty and smelling of smoke.

I walked over to a young family.

"How ya doin', where are you coming from, where are you headed?" I said as I neared them.

"Fort McKay, we've been there since it started. Well, we drove there because of the fire. We thought it would be safer to go north," a young woman with a baby on her hip told me. Her blonde hair was pulled back into a ponytail. She looked exhausted. "There's a few thousand up at Fort McKay. They let some of us come out through Fort McMurray early this morning. We left at five and drove right through town with the police. The fire is still burning in Fort Mac."

Many had evacuated to the north. Most were staying in residences at the oil sands facilities and at the community centre. Around 25,000 people had gone north on Highway 63 to the work camps around Fort McKay. Early that morning it had been deemed safe enough for vehicle convoys to be escorted by the police from Fort McKay south through Fort McMurray to safety.

"I don't know where we'll go, maybe back to Nova Scotia," another young mother said to me as her three-year-old son pulled on her T-shirt. "I just don't know when we'll get to go back. I don't know if our house is there anymore. I don't know what to do."

She started to cry. We hugged.

The day was getting warmer as the sun beat down on the dusty parking lot. Families sat on the grass eating hamburgers; so many were unsure what their next move would be.

"Do you know when it's going to be safe to go back?" a man asked.

"No, I don't. I don't think it will be this weekend or this week." My reply was like a gut-punch. He looked away from the camera and gazed into the unknown. Others were leaning against the side of their pickup trucks, looking up Highway 63 with a mix of hope and fear.

The flow of evacuees stopped as quickly as it began. That was it for today. The inversion had broken, the winds were stronger, and it was no longer safe to drive along the highway through Fort McMurray. We decided to make our way back to the barricade to look at the Beast. Shortly after 1 p.m., Shawn and I arrived at the parked procession of satellite trucks. The wind had picked up considerably. The smoke was acrid and thick, like fog; above us we could hear the water bombers and helicopters. Pyrocumulus clouds were forming overhead. These clouds form when the heat from a fire forces air, soot, and ash rapidly into the atmosphere. They look like thunderstorm clouds and can produce lightning and strong winds.

Dozens of cameras were set up on their tripods along the roadside. CBC, CTV, Global, and now NBC, ABC, CBS, and CNN were there; so were freelancers from the BBC, TV1 from France, ZDF from Germany, CCTV from China, and more.

The Fort McMurray fire was the biggest news story in the world.

The wind shifted slightly after three that afternoon. The smoke drifted to the north of the highway, and the air cleared overhead. About half a kilometre south, waterbombers dropped red fire retardant on an arm of the Beast, where the flames were reaching skyward. To the north of us, perhaps 300 metres off the highway, two helicopters were dropping buckets of water on a smaller fire.

The strong afternoon wind had curled a large plume of smoke around our position at the barricade on Highway 63; the sky and air were clear in our immediate area. Less than a kilometre away, the smoke coalesced into a massive cloud, and we could see the flames rising 10,000 metres up into the cloud structure along the horizon, eventually shrouding the sun from view. We had sore throats for a week due to the smoke.

The fire was having an impact on operations at the oil sands complexes north of Fort McMurray. Shell shut down production at Albian Sands, about 70 kilometres to the north. Suncor and Syncrude were also pulling back on operations at several of the largest sites in the industrial complex that lay across the oil sands—Millennium, Mildred Lake, and North Steepbank.

Every day a steady stream of 737s and Q400s flew back and forth from Edmonton and Calgary to the private airfields at the mine sites north of Fort McMurray. Thousands of oil workers were airlifted out so that evacuees could be accommodated in the camps' residences.

The production stoppage totalled a million barrels of oil a day, which is about a quarter of all oil production in Canada. The loss of output was forcing the pump price for petroleum products to rise across Canada, and it was costing the Albertan economy $70 million a day in lost revenue.

It was nearly four in the afternoon when three pickup trucks sped past our media location. The occupants sat three abreast in the cabs; the rear beds were filled with pet travel carriers. "Save the Animals" was spray-painted on one of the trucks. The RCMP waved their arms for the mini convoy to slow as it approached the barricade.

The trucks stopped in front of the police cruisers. Save the Animals wanted access to the city to look for and rescue abandoned pets.

The RCMP explained that it was too dangerous, and they did not have the resources to accompany them on this goodwill mission. The debate went on for several minutes, with each side restating its position.

Shawn pressed the record button and then hoisted the camera onto his shoulder.

"This is a part of the story that few of us had thought about," I said. "Over eighty thousand people fled this city on Tuesday. How many hadn't been able to find or take their pets? If faced with the choice of what to take in a crisis, what would you do? What if you couldn't find your cat? How long might you stay before the fire and heat forced you to go?"

As I talked, the group returned to their trucks; their mission would not go forward. They drove across the grassy median and, in a show of emotion, screeched rubber as they accelerated southward on Highway 63, retreating from the fire.

Shawn and I walked back to our Jeep.

"That was weird," he said.

"No kidding," I replied. "I wonder what is going on with the animals, though." We called the office to let them know that we would be sending along the report and explained what had just transpired.

"Weird," they said.

That night, back at the cabin, I received a call from Scott Meiklejohn, my producer in Oakville.

"Two things," he said. "First, I made some calls about the animals. From what we gather, the firefighters have been going house to house all week to look for stray pets. They are keeping them at a firehall in town."

"That's some good news," I replied.

"Second, can you do a Skype interview with *Good Morning Britain* at 2 a.m.?"

"Yup, no problem," I said. "Text me the details."

My phone woke me at 1 a.m. the next day. I showered and then checked the latest weather models. The fire was now nearly as large as metropolitan London, a city that burned for four days during the Great Fire in 1666. I also scrolled through local news sights in Alberta to find any interesting stories that I could relate on the Skype call.

In Calgary, a group of newly arrived Syrian refugees had given everything they had to fire evacuees. Wow, I thought. These people had fled a civil war, walking across Turkey or paddling the Mediterranean with nothing but the clothes on their backs only to live in refugee camps before coming to Canada, and they had given everything they could to Canada's first climate refugees.

After the interview I walked around Wandering River, taking a moment to process the disaster. I looked at my phone: #YMMStrong and #YMMFire were trending on Twitter and other social media sites. YMM is the airport code for Fort McMurray International Airport, which the fire had nearly consumed just two days ago. The men and women who were battling the flames had performed a miracle and kept the airport safe. In doing so, they ensured that the helicopters and aerial tankers were able to fly in and out of the closest airfield to the fire. Every single advantage helped in fighting the Beast.

The convoys of cars and trucks from Fort McKay began arriving around seven in the morning.

"We want our parents to know that we're okay, the kids are safe. We are going to drive back to Ontario, and we'll see you in a few days," one man said.

Our broadcast allowed people to let loved ones know they were going to be all right. It felt good to help connect people.

It also felt good to constantly relay the Red Cross numbers and web address. Tens of millions of dollars were raised on one weekend alone; nearly $320 million would make its way to the Red Cross in aid of Canada's first climate refugees.

Good progress had been made in pushing the fire away from Fort McMurray, and a fuel and resource point had been established on Highway 63 an hour north of Wandering River. That would be our next destination.

As we drove north, we saw several transport trucks heading to the oil sands industrial camps. Food, fuel, and all the other necessities that would have been so quickly depleted by evacuees over the past few days would soon be replenished.

Beside the southbound lane, not far from Mariana Lake, we saw the resource point. Several transport trucks, two fuel bowsers, and a collection of portable toilets were set up at a rest area beside the road. We drove across the grassy median to the makeshift site and parked the Jeep.

While we were prepping our gear, several cars arrived from the north. Families spilled out and headed to the toilets. A burley man, unshaven, wearing overalls and a dirty T-shirt, asked them if they needed gas.

"Yes, we do," they said.

The man climbed into the cab of one of the fuel bowsers and backed it to their car. He then climbed down from the cab, released a hose from the rig, opened their fuel cap, and began filling their tank. When he had finished, he screwed the gas cap back on, attached the hose to his truck, and waved at the kids, who were walking from the toilets with their mom and dad.

The father reached to his back pocket for his wallet.

"No need for that, sir," the man said. "There's water and food, sandwiches, you know, stuff like that, right over at that trailer." He gestured to one of the open transports. "Take whatever you need. There's lots of stuff. It's there for you."

The scene played out time and again—the generosity, the genuine caring for strangers, help for someone who needed a hand.

I asked the man if he would talk with me on television. He was taller than me, and his blue overalls were greasy from working the fuel truck. His hands were big and grimy. A cigarette was positioned behind his right ear, and there was a well-worn Edmonton Oilers ball cap stuffed in his rear pocket. He worked driving trucks from Edmonton to Fort McMurray, where he brought in groceries five days a week all year round. His home was just outside the capital, and his employer had volunteered his service and the equipment and supplies. He told me he didn't want to be paid for this work.

"It's what you do," he said. "I wish I could help more. It's so good to be an Albertan, to be a Canadian. This is what we do. We look after each other, no matter what."

He was looking at me earnestly. A tear rolled down his cheek, leaving a clean little trail through the dirt on his whiskers.

"Please, could I just have a hug?" he asked.

We all dropped a tear as we took a group hug. That moment allowed us to realize that no matter who we were, where we were, or what our role in life was, we were all human beings.

Over the weekend the prevailing wind shifted slightly and began pushing the fire northward toward Fort McKay but away from Fort McMurray. This change in the weather helped the firefighters gain control of the flames nearest the townsite.

On Monday afternoon, a helicopter flew over the media camp at the barricade. It came in low and flew north. A pair of motor coaches drove past us and stopped in front of the RCMP cruisers. The chopper returned and landed on the opposite side of the highway. The media were being taken on a tour of Fort McMurray.

The drive into town was quiet. A press officer explained that the tour would be quick; there would be a visit to two neighbourhoods impacted by the fire. Each stop would be short with an opportunity to leave the bus briefly, but we were not to leave the general area or touch or take anything, as it was still an active scene.

The bus drove through the community of Beacon Hill. Ash coated the ground like a dusting of snow. The skeletons of burned-out cars and twisted metal marked where houses had once stood. The air was still and smelled of burnt plastic; the acrid odour came from vinyl siding that had melted in the fire. The fine grey and white powder stirred as reporters walked along the street.

The fire had shown no deference in what it chose to consume. Some homes stood, the siding melted like creepy Halloween decorations; then there would be an entire block that was empty, just ash and recesses in the landscape where basements lay. Three thousand two hundred and forty-four homes, apartments, condos, offices, stores, businesses, and other structures had burned to the ground over a three-day period, most of them reduced to a fine grey ash.

We were taken next to the Abasand Heights neighbourhood. Heaps of smouldering timber and twisted steel were the only remnants of a block of apartments. Across the road, a building stood untouched. The randomness of the disaster was evident. After just ten minutes, we boarded the bus and were driven to the main firehall, where a briefing was conducted.

"When will residents be allowed to return home?" one reporter asked.

In very specific terms, Chief Darby Allen was able to underscore a simple fact: McMurryites would not come back until the community was safe from the threat of fire, and even then there would still be much to do before it would be deemed safe.

The entire water system had to be purged and cleaned, electricity needed to be restored, and in many cases entire systems of power lines would have to be restrung to homes. Natural gas and sewer lines would be checked for leaks and repaired. There was also the threat of contaminants, such as arsenic and various heavy metals, that were released into the environment with the combustion of so many common household materials. The fine ash that covered the townsite was likely toxic in many places.

The entire infrastructure and environment needed to be assessed, inspected, and cleared. At best it would be weeks, if not months, before people could return home.

MWF-009, the fire dubbed "the Beast," would continue to grow as it moved away from Fort McMurray. It wasn't brought under control until July, and would not be declared extinguished until early August.

It would consume 598,000 hectares of forest in Alberta and Saskatchewan, an area the size of the province of Prince Edward Island. The damage was estimated to be $9.9 billion, making it the most costly Canadian disaster. The evacuation was the largest in Alberta history. The oil sands facilities would take several months to resume full production. The first families began a phased return to Fort McMurray in June. Slowly, throughout the summer and autumn, they would come back in groups.

Miraculously, the fire took no lives.

CHAPTER 9

Gators

Florida, 2016

In early October 2016, I was sitting at Pearson Airport in Toronto with my producer, Rod Heinz. We were going to Florida to create a series of high-quality infomercials to draw Canadians south to the state that coming winter.

It was an interesting time to visit the United States. The election was a month away, and it seemed that history was about to be made again, with Hillary Clinton a near shoe-in to become the first woman elected to the office of President of the United States of America. Donald Trump, the flamboyant real estate developer and reality television host, had been a brash and disruptive opponent, with a platform that seemed to be based on rage and outlandish behaviour. Our trip would include a visit to one of his golf courses, and I felt it would be interesting to gauge the temperament of our southern neighbour.

It was Sunday, October 2. I'd finished my weekend show at around noon and met up with Rod in the equipment room at the office. We packed up cameras, a Dejero unit, batteries, microphones, cables, tripods, lights, a couple of computers, and all the other equipment we needed to shoot the series.

Everything was packed into the big, black, nearly inde-structible travel cases that pretty much double the weight of the equipment. Between us, we had five bags each. We split the load between our cars and drove to the airport. Customs would be our first stop. All the equipment would be searched, docu-mented, and verified against a list that had been emailed to U.S. authorities earlier in the week. The inspection took about two hours. After getting the right stamps and approvals, we left with all our gear packed precariously onto two large carts and went to the check-in line. We paid the overweight and extra baggage fees, then headed off for U.S. Customs. We'd been at the airport for nearly three hours. This is what it's like travelling as part of a TV crew.

Once we cleared customs, we headed for our gate. We arrived in time to order a sandwich and began discussing our production plan for the next few days. We also talked about Hurricane Matthew.

For several days, Matthew had been churning across the Caribbean; it was now about forty-eight hours away from Haiti and could become a problem for Florida later that week. We decided that we would need to accelerate our filming schedule to stay ahead of any foul weather. It was already four in the afternoon. I'd been up since shortly after two in the morning, and we wouldn't be at our hotel until at least ten that evening, if all went well.

We finished our sandwiches as the last boarding call for our flight came. Rod and I walked quickly to the gate and had lined up behind a half-dozen people when a familiar face breezed by and walked right up to the agent at the business-class check-in. Rod and I looked at each other: *That's Peter Mansbridge.*

Peter was the host of CBC's *The National*, Canada's popular nightly newscast. He was, and in my opinion remains, one of the most trustworthy and reliable news presenters in the world.

The Weather Network had an agreement with the CBC: we provided almost all their weather services on the main network and on CBC News Network. So I'd been presenting the weather on *The National* several times a week. Our weather segment would air toward the end of the newscast, unless there was a major weather event, in which case the story was moved up. The weather segments were recorded about an hour prior to the actual airing of *The National* and were delivered by a digital link from our Oakville studios.

So I had never actually met Peter in person.

He was getting settled into his seat and had a nice tomato juice in front of him when Rod and I shuffled by with our bags of lithium-ion camera batteries and the Dejero unit on our way into economy. For the next three and a half hours, I sat directly behind Mr. Mansbridge, a curtain separating the news from the weather.

Our flight landed on time in Fort Lauderdale, and when we disembarked I caught up with Peter and introduced myself. He was in Florida for a speaking engagement, but he was very interested in the hurricane and we discussed the potential for tragedy in Haiti.

Haiti occupies the western half of Hispaniola, the second-largest island in the Caribbean Sea. A major geological fault line between the Caribbean and North American tectonic plates runs parallel to Haiti. The Caribbean plate, which accounts for the area south of Cuba, is sliding slowly to the northwest at about 10 millimetres per year. Occasionally, for a brief moment in geologic time, that movement accelerates. Those sudden bursts of movement result in earthquakes. In 2010, it was estimated that over a quarter million people died in a magnitude 7 earthquake in Haiti. Since then, there have been even stronger episodes of seismic activity and, sadly, more deaths.

In addition, Haiti and the islands of the Greater Antilles are also along one of the most frequented paths for hurricanes during the Atlantic storm season. The coming days would prove to be devastating, and not just to that country.

The baggage carousel was spitting out suitcases, baby car seats, and boxes. Passengers retrieved their baggage until only Rod and I remained, waiting for the last two of our equipment cases. Then the carousel stopped.

We went to the lost luggage counter and learned that the missing equipment cases had been left behind in Toronto. We began the process of filling out several pages of paperwork to get a delivery arranged sometime the next day.

The next load of passengers began arriving to claim their bags. This group was inbound from Montreal and included Nathan Coleman, our Atlantic Canada video journalist. I'd covered more weather events with Nate than anyone else. He's a superb cameraman, or "shooter" in broadcast lingo, and his easygoing, up-for-anything personality made him a lot of fun.

By the time we picked up our rental, a large Suburban SUV, and navigated our way to the Crowne Plaza on South Ocean Drive, it was nearly midnight.

Meanwhile, Hurricane Matthew had been deemed a Category 4 storm. Category 4 means, among other things, that wind speeds are up to 251 kilometres per hour. Matthew had rapidly accelerated in speed and taken a turn northward, one that would place it over Haiti in just twenty-four hours. The question people were asking was whether Matthew would track northward to cross the Bahamas 100 kilometres to our east or veer westward toward us, in South Florida.

The next morning, at around four-thirty, the familiar ringtone of my iPhone woke me. It was Derek Snider, executive producer of The Weather Network.

"Hey, listen, sorry to wake you up. Would it be possible for you guys to shoot a couple of pieces from the beach this morning about Hurricane Matthew?" he asked. "It went to Category 5 strength overnight, and it's heading for Haiti today."

Derek had come to The Weather Network with a rich and varied background in sports broadcasting. His ability to liken our presentation of big weather events to big sporting events was unique and really quite brilliant. "This storm will be our Stanley Cup," he was fond of saying. Those words motivated people to work harder, dig deeper, and find the great stories within the story.

"Why don't we go live with the morning team starting at six," I said. "It's doable and not a problem on this end."

He thought that was a great idea and would get our network well ahead of the curve with feet on the ground for this developing story. All I had to do was wake up Rod and Nate.

The three of us met in the hotel lobby at 5:30 a.m. Nate had his camera gear, Rod brought the Dejero, and I offered apologies for the early call. We decided to shoot our material across the road on the beach. It was still dark, but the sun was beginning to rise and the eastern sky over the Atlantic was pink in the morning light.

We spent the morning on the beach, sending reports about Hurricane Matthew to Canada and recording our vignettes to attract tourists to South Florida. As we spoke about winter holidays and warm ocean water, families played in the sand and surf behind us. It was surreal, what might happen here between now and then.

We finished filming and walked across the road to the hotel. Our last equipment cases had been delivered and were waiting for us in the lobby.

We had a busy schedule planned. An afternoon conference call to our meteorology and television production team would determine our priorities over the next few days. The consensus was that Hurricane Matthew would cross eastern Cuba and the western Bahamas in the next forty-eight hours. Computer models and our best forecasters felt that the storm would then run parallel to the coast of Florida, with a strong chance of coming onshore. If it made landfall, it would happen somewhere between Miami and Jacksonville.

It was estimated that nearly a million Canadians spent a significant portion of the winter as snowbirds. Nearly half that number owned property in the state and would want to keep track of any possible damage. The Weather Network and our French-language service, MétéoMédia, would each send a reporter and camera operator to cover this increasingly important storm. Both teams would be in Orlando the next afternoon. Now there would be three teams spread across the state, so that one of us would be near any potential landfall.

That evening we shot from the Little Havana neighbourhood in Miami. Cafés were filled with diners, and music carried into the sultry night air. The locals, mainly Cuban expats, were concerned about both the election and the hurricane.

I talked with two men who were playing dominoes and smoking cigars in front of a quaint family restaurant; they had strong political opinions regarding each candidate, but mostly against the present, long-standing Cuban regime that had forced them to flee their homeland in a boat across the Straits of Florida. They were more concerned about family they hadn't seen in thirty-five years and the impact Hurricane Matthew might have on them. The storm was bearing down on Haiti and eastern Cuba, and we'd be up early to file new reports in the morning.

That night, Hurricane Matthew crossed over the Massif de la Hotte on the Tiburon Peninsula of Haiti, nearly obliterating communities along the northern coast. It was still a Category 4 storm, the strongest to impact Haiti since Hurricane Cleo in 1964. It had also been just six years since the country's last devastating earthquake. That quake had struck just before 5 p.m. on January 12, 2010. The magnitude was measured at 7 on the open-ended Richter scale. Officially, 200,000 people died, although international relief agencies said the number was more likely 250,000. The damage to infrastructure had been estimated to be nearly US$8.5 billion.

Now, in Port au Prince, the country's capital city, the wind speed measured 172 kilometres per hour. Across the border in the Dominican Republic, more than 200 millimetres of rain had fallen overnight. Officially, 546 people had died in Haiti, but relief agencies suggested the number was likely in excess of a thousand victims. The damage to infrastructure and homes was estimated at nearly US$3 billion. A million and a half people needed urgent assistance.

We reported the grim news throughout the morning. In the afternoon, we continued recording our promotional project, this time at the PGA National course, which was owned by Donald Trump. The facility was opulent, and every detail of our visit was arranged to perfection. Our handlers had answers to all of our questions, the catering for lunch was five-star, and Donald Trump was always referred to as "Mr. Trump."

The chasm between this reality and the reality unfolding on the Caribbean islands began to temper our thoughts. All three of us had become reflective as we finished our work and set out on the four-hour drive from West Palm Beach, across the Everglades, to Naples, on the west coast. Near Miami, we pulled into a service station and joined a long line to fill the Suburban with gas. People were preparing for Matthew. Grocery stores were beginning to run out of bottled water, and shop owners had started securing plywood sheets over the windows of their stores.

We made it to Naples by sunset. Traffic along the I-75 had been heavier than expected; it was apparent that many people who lived in the Miami–Fort Lauderdale area had decided that they would wait out the storm on the west coast. Typically, hurricanes that run along the Atlantic coast of Florida have minimal impact on the state's Gulf Coast communities. There is, however, a significant increase in hotel bookings as east-coast residents head west to escape the worst of the storm.

I didn't sleep well that night. Thunderstorms rolled through most of the evening and early overnight hours, rain came in a torrential downpour, and the sky flashed like a strobe light. I read reports of the damage that Matthew had caused in eastern Cuba.

The next day, we rose well before sunrise to set up our

camera and lights in Tin City, a small, historic part of downtown Naples. The buildings along the Intercoastal Waterway are made mostly of corrugated steel and aluminum and look the way many old Florida Gulf Coast towns must have looked in the mid-1800s. We chose the location to underscore the types of structures that people in Haiti and even Cuba had lost to the hurricane. Tin City also had a marina filled with million-dollar yachts.

We reported on the estimated US$2 billion in damage Hurricane Matthew had done in Cuba. More than a million Canadians visit Cuba every winter, a number that has been consistent for nearly a decade. We also reported on the latest news from Haiti; nearly 200,000 people were homeless. Initial damage estimates were beginning to indicate that this storm had been devastating. Ultimately, Matthew would leave over US$2.8 billion in damage, making it the most expensive in that nation's history. We wrapped up at nine in the morning and headed east to the Everglades for our sales project.

The Everglades are a vast wetland that covers the southern third of Florida. The area begins not far from Orlando in Central Florida at the Kissimmee River and flows into Lake Okeechobee, which is the second-largest freshwater lake in the lower forty-eight states and has an average depth of only 2.7 metres. At the southern end of the lake, water flows like a great river of swamp and reedy grass, 100 kilometres wide and stretching 160 kilometres south to merge with the mangroves that line Florida Bay. The Everglades are a unique ecosystem, one unlike anywhere else on Earth.

In the 1940s and 1950s, 2,300 kilometres of levees, canals, and dams were built to divert water from the Everglades for agriculture and to supply the thirsty and rapidly growing cities of South Florida. Over the next three decades, the wetland shrank by 50 percent, causing tremendous ecological damage. By the 1980s, efforts were underway to rehabilitate what was left of the Everglades and end further degradation to the ecosystem. It is now a UNESCO protected site.

Rod, Nate, and I sped through the sawgrass and along sloughs deep into the wilderness with Carl, the gladesman at the helm of our airboat. He wore jeans, construction boots, a T-shirt, and a twenty-year-old ball cap, and sported a long red beard. There was a shotgun at his side and a Ruger SR1911 on his hip.

The Everglades airboat is a flat-bottomed, open-air vessel that draws less than three inches of water. A V8 engine is mounted on the stern, driving a large airplane propeller that pushes you along at 80 kilometres per hour. Your seat for this epic ride is a bench, level with the water, the hull rising only a foot above the waterline.

There we were, three city slickers with all our fancy cameras roaring deep into gator country with a well-armed man we had just met. We learned that Carl had lived all his forty years in this one patch of Florida. Born in Ochopee, he lived "back in the swamp" and had driven airboats all his life. He went into Walmart, in Naples, he said, "about once a month to get some groceries and to stock up on ammo." I wanted to ask why he went through so much ammo but decided not to.

We spent that clear, hot afternoon exploring the glades. Carl spoke about the ecosystems, the history, the storms, and his thoughts on the uniqueness of this natural wonder. He was a gifted storyteller and knew the science, the lore, and the traditional Seminole history and legends of the Everglades.

This was Carl's home, and he loved its history and its natural beauty.

"The Seminole People, they ran the other Underground Railway, you know," Carl explained. "They helped the slaves who ran from the plantations, kept them hidden here in the glades, and then helped them sail to Andros Island." Thousands of slaves had moved through Florida to find freedom in the Bahamas in the 1820s.

Carl was well read on climate change, too. His voice wavered slightly as he talked about the coming devastating implications for the Everglades, but equally for Belize, the Maldives, and other low-lying regions.

"The mangroves are an invasive species here," he said. "They're overtaking the natural coastline where the freshwater flows into the ocean. The rising sea levels will change this ecosystem."

Carl knew that as more salt water seeped into the Everglades, the natural grasses would die and be replaced by the mangroves, which thrive in partly saline water. He spoke about how we all needed to better look after our planet, to respect nature and follow its cues.

As we weaved around cypress stands, we learned about manatees and were introduced to both alligators, which live in the freshwater channels, and crocodiles, which live in the mangroves in salt water closer to the coast. As we drew into a narrow channel, Carl cut the motor and the airboat drifted forward silently. You could hear the songs of hundreds of birds. Carl rose and moved around us to the bow. He carried a broom handle and leaned over the prow, peering into the water.

"Liza should be around here this time of day," he said. "I

want you to meet her." Liza turned out to be a 4-metre-long alligator. "She can be mean," Carl said as he poked at the water with the broomstick.

Liza appeared—or rather, lunged—out of the brackish water with a splash. She resurfaced slowly, her head and back just above the water, her mouth opened for us to see her teeth.

"She's mean because she lost an eye fighting another gator," Carl told us. He went on to explain that it is extremely rare for an alligator to attack or eat a human; we adults are simply too large. "But if one should ever charge you, run as fast as you can, directly away from it," he said. "You can out-run one. They'll grow weary of a land chase pretty quickly."

Good to know.

We pulled up to the pier in Ochopee. We filmed a lot of material, enough to create a half-dozen tourism vignettes. A three-hour drive back to Fort Lauderdale lay ahead of us; Matthew was due to make landfall near Cape Canaveral within the next twenty-four hours.

As we drove east on Tamiami Trail, we saw a steady stream of traffic heading west. Families packed into minivans and cars were heading for Naples and Fort Myers. The progression of vehicles added to our sense of foreboding as we made our way back to the city.

The mountains of Haiti and Cuba had weakened Matthew's air-circulation patterns. A hurricane draws its energy from the heat stored in warm ocean water; passage over land disrupts that supply, and steep mountain terrain disturbs the cyclonic circulation near the surface. The storm had decreased to Category 3 strength, but Matthew was still a major hurricane, with winds upward of 207 kilometres per hour. Storm surge, which is the upwelling of the sea driven by wind and low atmospheric pressure, could lift the water 4 metres above its normal level.

In the early hours of the next morning, Matthew began passing over the Bahamas. The hurricane had gathered energy from the warm waters of the Atlantic Ocean and was back at Category 4. There would be no mountains in the way of the storm now, just warm water to feed energy into the atmosphere.

It was overcast and windy as we began our live coverage at six o'clock that morning. We broadcast at the end of Hallandale Beach Boulevard. The surf—large 3- and 4-metre breakers—crashed onto the shoreline behind us. During the four hours we spent broadcasting from shore, the growing surf and powerful rip currents ate away at the beach. Millions of tons of sand were being drawn back into the sea.

These same waves were flooding Andros Island in the Bahamas. The storm was racing to the north-northwest. Nassau's airport had reported winds of 233 kilometres per hour. These field reports, combined with the latest computer modelling for the next twelve hours, prompted a flurry of public safety measures.

We did a live shot with The Weather Network and the CBC, explaining the latest developments. Behind us on Hallandale Beach Boulevard the lift bridge across the Intercoastal Waterway was being raised. Police cars guarded the approach to the bridge. All access to the coastal islands was now prohibited.

We were on the island and would ride out the worst of this storm at our hotel. Our rooms were on the eighth floor facing the ocean to the east, and we would report live from the balcony when it became too dangerous on the ground.

To the north, our other reporting teams were holed up in concrete parking garages on the barrier islands near Cape Canaveral. Later, when cellphone towers were toppled by Matthew, they would lose the ability to communicate with the outside world. Their last report came in the late evening. We wouldn't hear from them again until the next afternoon.

The Kennedy Space Center, Disney World, and all the other Central Florida attractions were closed. "Shelter in place" was the message from the police and local and state governments.

By late afternoon, Matthew was gusting with winds of nearly 200 kilometres per hour, the rain was driven sideways, and palm fronds and debris were lifted skyward and flew past us on our eighth-floor perch. As we reported live to Canada, the eye of Hurricane Matthew was exactly 120 kilometres east of us. That evening the storm moved along the coast, just barely offshore. Twice the eyewall of Matthew came close to a landfall, once near Vero Beach and again late that night near Cape Canaveral.

It is surprising how fast a hurricane passes. For about ten hours, Rod, Nathan, and I experienced what I would consider to be stormy weather. Matthew built to its crescendo between noon and 6 p.m. that day. By eight, we were walking the streets surveying the damage. The worst was over in this part of Florida.

The wind was abating. The rain came as light intermittent showers. The night was balmy and quiet. The streets were wet and deserted; some were flooded, but mostly they were strewn with palm fronds and coconuts. Traffic lights flashed orange and the Hallandale Beach Boulevard Bridge had been lowered. The police cars were gone. We came across several homeless people who were going through trash bins, their wet possessions in grocery carts they slowly pushed up the middle of North Ocean Drive. This was the aftermath of Hurricane Matthew in Broward County, Florida.

The three of us were up at five the next morning. We would drive north along the I-95 toward Port St. Lucie and then get on the A1A, continuing north to document the damage on the barrier islands and in cities like Jupiter and Fort Pierce. We rigged the camera and Dejero unit inside the Suburban so that we could broadcast live from the truck en route. Nate ran the camera and communications link to The Weather Network from the back seat, Rod drove, and I sat in the passenger seat to report on the latest developments in Florida.

Hurricane Matthew was abeam Neptune Beach at high tide, and there was significant storm-surge flooding in both St. Augustine and Jacksonville. Nearly a million people were without electricity. As we broadcast from the Suburban, convoys of trucks passed us, their amber lights flashing in the early morning light. Power company vehicles and teams of lineworkers were heading north to begin their work.

The median and ditches on both sides of the interstate were filled with water. At times the pools linked across the highway, forming ponds. Near the town of Stuart, we pulled off the road to provide an update for the CBC.

I waded into water that filled the median; it rose to just above my waist. The bottom was soft and muddy. As I reported on the aftermath of Matthew, I began thinking about alligators. Was it possible that one might have been displaced by the storm and was, right now, sharing this median with me? Carl's wise words about gators weren't lost on me.

Hurricane Matthew stayed offshore, which spared Florida significant damage and loss of life. By the afternoon, storm surge was flooding the low-lying coast of Georgia. The storm had slightly weakened—the wind was down to 175 kilometres per hour—but the hurricane was driving a wall of seawater inland.

Much of Savannah was inundated. It is more difficult to evacuate cities like Savannah and Charleston—there is one main artery inland and away from the hurricane. Usually, all four lanes are opened to traffic in just one direction to help move the volume of evacuees from harm's way. Such was the case today; tens of thousands were on the move.

As that drama unfolded more than 400 kilometres north of us, we continued to survey damage along the Florida coast. Our drive along the A1A on Hutchinson Island offered a glimpse of the work that lay ahead to restore the electricity. For miles, the power poles were down, some beside the road, others across it. We weaved around the tangled mess of wires that littered the route. Occasionally an oddity would surprise us, like a boat in the middle of the highway. There was no traffic.

We came upon a house where several Florida pines had been felled. It was very likely that the previous night the wind on this exposed island had gusted at 250 kilometres per hour. The trees had crushed the sunroom at the front of the home but also collapsed the adjoining cinderblock garage. We used this scene as a backdrop to illustrate the type of damage that many Floridians were facing that day. Nathan, Rod, and I ran through how we would shoot the report. The shot would start tight on the tree roots, as I explained that the sandy soil composition on the barrier islands wasn't ideal for trees trying to withstand hurricane-force winds. We would then walk alongside the fallen tree and end in front of the damaged home.

We had moved into position and were waiting to go live on The Weather Network when we heard two pulses of a siren, *woop, woop*. All three of us looked toward the road. A sheriff was stepping out of his black-and-white cruiser, lights flashing. He was a big man with a crew cut, standing tall in his crisp uniform with his hands on his hips.

"Boys, y'all right now are breaking the law. This is private property, and what y'all are doing is illegal and called trespassing," the sheriff said, each word delivered deliberately with a slow Southern drawl. "A person would be well within their rights to go ahead and shoot you right here, on account of you breaking the law today by being trespassers on private property."

Rod, Nate (who had a camera on his shoulder), and I looked at each other, then at the sheriff, then at each other again.

"Hi there," Rod started. "We're from a TV station in Canada reporting on the hurricane."

Nate and I nodded in agreement.

The sheriff looked at us for a moment, then turned his head and looked out over the Atlantic Ocean. Finally, he turned back to us and said, "Y'all need to gather your stuff and get out of here, now." He then opened the cruiser door and slid back into the driver's seat.

As the sheriff closed his door, my IFB came alive with the voice of our producer back at The Weather Network in Oakville. An IFB is an earpiece connected to a small transmitter that allows you to communicate with the control room.

Through it I can hear instructions from my producer and questions from the on-air hosts.

"We're going to take you live in fifteen seconds," she said.

Nate was wearing an IFB too. We looked at each other as the sheriff slowly pulled away and headed down the road.

"Let's do it really fast," I said to Nate.

We quickly positioned ourselves in front of the fallen tree as the producer counted us in: "Four, three, two, one, and on you." As we ran through the details of the story, Rod started the Suburban, pulled it out of the driveway, and opened the rear door. As soon as the producer said, "You're clear," Nathan and I grabbed our gear and hopped into the truck. We looked at each other and shook our heads. It was time to wrap this day up.

We drove back to the hotel in Fort Lauderdale, listening to Bruce Springsteen on satellite radio. We didn't talk much. We were tired from the lack of sleep, tired of Matthew, and our minds were wearied by thinking about how this weather had such varied impacts on so many people.

There were the families who packed their cars and left town to spend a day or two away from the storm in a hotel, having dinner at Pizza Hut and hoping their house would be okay. There were those in Haiti who'd lost their crops, their homes, and their livelihoods. After waiting for four days for basics like clean drinking water, help for them was only now arriving. A humanitarian crisis was underway there again.

Matthew would make landfall the next day near Charleston, South Carolina. The flooding left Wilmington under a half-metre of water, and 120,000 trees were brought down by the wind across the Carolinas. The post-tropical remnants of the hurricane brought over 200 millimetres of rain to

Atlantic Canada, causing widespread flooding. Sydney, Nova Scotia, broke a record for rain—225 millimetres in one day. Throughout that province, basements flooded, and 50,000 customers lost power, many for several days. The wind had gusted at more than 100 kilometres per hour over Thanksgiving weekend. In Newfoundland, entire communities, such as St. Alban's, were cut off when torrents of floodwater washed away roads and bridges. In Canada the economic losses from the storm came in at around $150 million—a big number, but tiny when measured against the losses in the Caribbean.

Hurricane Matthew broke many records on its long destructive journey. At the time, it was the longest-lasting Category 4 storm ever recorded during the month of October, holding that strength for four straight days. Matthew also intensified from a tropical storm to a major Category 5 hurricane in just twenty-four hours. It gathered more ACE, or accumulated cyclone energy, than any other storm in the eastern Caribbean, and was one of the highest energy generating storms of the past fifty years.

Due to the extensive damage that it caused along its track and the loss of life incurred, the World Meteorological Organization retired the name Matthew.

CHAPTER 10

Snowbirds

Florida, 2017

The rain blasted sideways, driven by steady winds of over 200 kilometres per hour. Metal trash bins, sheets of aluminum, tree limbs, lawn furniture, and other deadly projectiles filled the air. Storm chasers huddled inside the concrete stairwell of a parking garage. It was too dangerous to be outside.

Hurricane Katrina was making landfall in Buras, Louisiana.

Katrina went from a Category 3 to a Category 5 hurricane in just nine hours, making it, at the time in August 2005, the strongest storm ever recorded on the U.S. Gulf Coast. The levees that protected New Orleans were breeched, and the city, which lies at or below sea level in some areas, was instantly flooded.

In the aftermath, 1,800 or more lives were lost. It was the most costly natural disaster in U.S. history, with US$182 billion in damage. The federal response to the disaster was widely criticized, and there was significant environmental damage due to the large number of petroleum and chemical facilities in the area.

We are fortunate in Canada. Our location as a northern

nation helps ensure that when a hurricane impacts us, it is usually in its decaying stages, and its most devastating effects generally lie behind it in the far-off Caribbean or along the coast of the United States.

But we do make up a large segment of the population in many of these warmer climates. More than half a million homes in Florida are owned by Canadians, and we make millions of trips to the U.S. Gulf Coast, Florida, Mexico, and the sun-drenched islands of Cuba, Dominican Republic, Puerto Rico, Bahamas, and countless others every year. We have a vested interest in the people we meet and get to know in these places; we enjoy the natural beauty and climate of these regions. For a time there was even a strong movement to make Turks and Caicos our eleventh province.

We know these areas well; we also understand the power of these storms. All of Atlantic Canada has been hit by hurricanes, and many have caused significant damage and death. Even Toronto, in the heart of the Great Lakes, bore the brunt in October 1954 of Hurricane Hazel, which resulted in eighty-one deaths in Ontario.

Stronger, larger, and more frequent hurricanes are becoming the new normal. In 2017 a series of large-scale hurricanes wreaked havoc on hundreds of millions of people over a period of just three months.

Hurricane Harvey developed off the coast of Africa on August 13, 2017, and tracked westward to the Caribbean, where it evolved into a tropical storm just four days later. When Harvey reached the Gulf of Mexico on August 22, the unusually warm water helped it gain strength, and a strong high-pressure ridge over the central United States kept the storm over the Texas coast for several days in late August.

Rain fell in torrential downpours as Harvey literally inched toward the petroleum-refining heart of America. For days the massive oil refineries that line the coast and supply half of all petroleum and natural gas production on the continent ceased production as their crews evacuated to safety.

Harvey crept along the Texas coastline from Corpus Christi to Houston and then to Louisiana, making landfall three times over six days. Houston is the fourth most populous city in the United States, with a metro population of over seven million. Half of the entire population was evacuated from the area as a record-breaking 1,550 millimetres, or 1.5 *metres*, of rain fell, and 150,000 homes were flooded. That amount of rain, about 100 trillion litres, would fill the Houston Astrodome stadium with water an astounding 85,000 times. Just half that amount would have been enough to end the drought in California.

The damage brought on by Harvey rivalled that of Katrina in New Orleans. This time, however, the crisis was better managed, and the human toll was considerably reduced.

At the same time, on August 13, Hurricane Irma was born near the Cape Verde islands, about 500 kilometres off the coast of Africa. Less than forty-eight hours after its incubation, Irma was a Category 3 hurricane, with winds fluctuating between 178 and 208 kilometres per hour. By September 6, Irma was the strongest storm on Earth, with sustained winds measuring at a steady 285 kilometres per hour, making it a Category 5 hurricane.

At dawn on September 6, Irma passed directly over Barbuda, destroying nearly every structure and tree on the island. Thankfully, the entire population had been evacuated to Antigua, 30 kilometres away.

By early afternoon, it would also pass over Sint Maarten and Virgin Gorda. The winds would lift vehicles from the roads and throw them, battered, in ditches. Seventy percent of the buildings on Sint Maarten became uninhabitable. The damage inflicted on the BVI, including Virgin Gorda, was estimated to be in the area of US$9 billion. It would take five months to fully restore the electricity.

At dusk, the hurricane skirted just north of the U.S. Virgin Islands. A significant browning of the islands took place there—a browning is when all the foliage is stripped from any tree left standing.

By late evening, the eye of Irma was 100 kilometres north of San Juan, Puerto Rico. The storm had a diameter of nearly 500 kilometres. Its most damaging winds reached more than 100 kilometres from its centre. Because the strongest and most devastating winds tend to be on the northern side of a hurricane, Puerto Rico was spared the level of carnage that was sustained in the Virgin Islands. Still, the island incurred US$1 billion in damage. Seventy-five percent of Puerto Ricans lost electricity, and half of all the cellphone and communications towers were destroyed. Over 360,000 residents lost water service and over a quarter of all crops were wiped out.

On September 7, Irma galloped along the northern coast of the Dominican Republic and Haiti. Again, the strongest winds remained offshore of Hispaniola, but the metre of rain that fell on the island led to widespread flooding and deaths.

By avoiding landfall and the mountains of Hispaniola, Irma had done herself a favour—the structure of the storm was not disturbed by the mountainous terrain of the island. Mountains interrupt the circulation patterns within a storm; a change to those patterns can decrease the strength of the wind. Irma was still drawing energy from the sea as it moved toward the northern coast of Cuba late that evening.

In Florida, a state of emergency had been in place for three days, and an evacuation of some 6.5 million residents in southern Florida was underway. It would be the largest evacuation in the history of the state.

Mark Robinson and I were headed south to provide coverage on Hurricane Irma, for both The Weather Network and CBC News Network. This was becoming a huge weather event and making headlines on both sides of the border. But flights had been cancelled and airports throughout the state were shutting down. The government did not want more potential casualties arriving by air. We found out that Air Canada was running rescue flights to Fort Lauderdale that evening. I asked the ticket agent if we could get on one of those planes.

"I can get you to Fort Lauderdale, but we don't know when the airports will reopen," she said. "So it's a one-way ticket. It's up to you."

"Book us, please," I replied.

It was odd to look down the long aisle and see just twelve people on a plane that easily accommodated over one hundred and eighty.

The captain came on the intercom to welcome us aboard and outline our route: south to Wilmington, North Carolina, and then along the Atlantic Coast to Florida. It was 9:30 p.m. I did the calculations in my head; we'd be in Fort Lauderdale at one in the morning.

I used an app on my phone to book a car rental in Fort Lauderdale. Then I began booking hotel rooms, in Fort

Lauderdale that night and Miami for the following three nights. As a precaution, I also booked rooms on the Gulf Coast for the same three nights as Miami, just in case the path of the hurricane changed. The west coast rooms were much more difficult to find, likely due to the influx of evacuees, but I was able to locate a couple in Sarasota.

Having covered dozens of storms, I knew how important it was to have a base to operate from. Once you get wet, whether from rain or snow, you tend to stay wet. You will also inevitably get everything around you wet too, including the inside of your vehicle and all of your equipment. A place to dry everything off is important when you are working thousands of kilometres from the office.

It was now ten o'clock Thursday night. We would be going live on Fort Lauderdale Beach in just eight hours.

Three hours later, the familiar sound of the Airbus engines rolling back to idle thrust, signalling that we were beginning our descent. I looked out the window into the blackness of night. Occasional lightning flashes illuminated the thunderclouds over the Atlantic Ocean. We were abeam Vero Beach, about 120 kilometres from Fort Lauderdale. The plane would be on the ground in twenty minutes.

I moved across the aisle to look out the starboard windows as we passed through some wispy cirrus clouds. The orange lights below outlined the urban areas of Port St. Lucie, then Jupiter and West Palm Beach, as the plane drifted down toward Fort Lauderdale.

Peering into the night, I could see three parallel lines of red lights extending northward along the coast—tens of thousands of cars, trucks, SUVs, and minivans, each filled with men, women, and children fleeing the impending storm. I had seen these evacuations from the ground. From the air, it was far more dramatic.

As the plane positioned for our landing, the rows of vehicles on the highways came into focus. Traffic was just creeping along. I thought about what those evacuees were feeling. This was a slow, methodical egress, not the panicked run-for-your-life flight we had seen in Fort McMurray.

As we walked up the jetway, it smelled like Florida—that humid, musty, and slightly floral scent. The gate was filled with hundreds of people; exhaustion and anxiety were etched on their faces.

We checked into the hotel at 3 a.m. and my phone buzzed me awake at five. Mark and I met in the lobby twenty minutes later. We stepped out of the hotel's air-conditioning and into the tropical air. We stopped for a moment to adjust to the heat, then headed east, carrying the camera and Dejero broadcast box two short blocks to the corner of SE 17th Street, the A1A, and Eisenhower Boulevard.

It was nearing 6 a.m. and already the air was thick with humidity. The temperature was 30 degrees Celsius and the humidex 36. Our shirts were soaked with perspiration.

A four-car police blockade at the intersection prevented us from crossing the bridge to access the barrier island that is home to Fort Lauderdale Beach. I walked toward a group of officers huddled in front of a cruiser. Early morning twilight was beginning to break.

I raised my hands above my head and said, "Excuse me, can we cross the bridge?"

The reply was quick and terse. "No, no access."

My earpiece came alive with the audio feed from The Weather Network from Oakville. "We'll go live with you and

Mark on camera in thirty seconds," our producer, Jamie, said. "Give us a recap of the evacuation process and when you expect the weather to get bad in Florida."

The eastern horizon was taking on the first hues of an orange sunrise. Mark adjusted the camera settings and then hurried over to stand beside me. We would begin our live broadcast to Canada from this intersection; the flashing red and blue lights of the police cars would serve as our backdrop.

"Storm Centre on The Weather Network," the recorded voice boomed in my earpiece, with the appropriate musical coda as accompaniment. The flashy animation for Storm Centre transitioned to a split-screen shot, with Mark and I on one screen, flashing blue and red lights pulsing behind us, and Chris Murphy at our Oakville studio on the other. We talked about the ongoing evacuation of the southern third of Florida and the damage Irma had done in the Caribbean the day before.

While we chatted, video of the hurricane passing over Sint Maarten played on screen. The scene was captured from a hotel balcony. The windows had been blown in, and there was debris flying through the air—palm fronds, furniture, strips of metal siding.

As the live segment ended, a police officer walked toward us. He was in his thirties, tall, with a crew cut; he looked like he was from central casting for a 1950s surf movie.

"What are you doing?" he asked.

We explained that we were from a Canadian television network, here to cover the hurricane.

Another officer suggested that we should broadcast from the beach. "There are a couple of nice places you could go. We'll give you an escort over the bridge. Go get your vehicle."

Water from the Red River forms a lake nearly 2,000 square kilometres in size over southern Manitoba. In this aerial image, taken in April 1997 near Morris, 10-metre-high sandbag levees are all that protect homesteads from being inundated by spring flooding.

Photo © The Weather Network

Encased in ice after days of freezing rain. Quebeckers suffered catastrophic damage to their power grid. As a result, millions of people spent days, even weeks, without electricity or heat. The deadly storm kept 20 percent of Canada's workforce from their jobs, and caused $5 billion in damage.

Photo © Christopher Moore/Corbis

Hurricane Juan and all the other storms the region has experienced are memorialized in wry artistic sculptures along the Halifax waterfront. The strength of the wind left ships floundering and even destroyed the instruments designed to measure its speed.

Photo © Chris St. Clair

Four and a half months after Juan, hurricane-force wind gusts and the largest snowfall in the history of Halifax, Nova Scotia's capital, created near-zero visibility. A state of emergency was declared, and a curfew was imposed to ensure public safety and allow plows to clear the snow. The storm was dubbed "White Juan."

Photo © The Weather Network

Some locations measured over a metre of snow. Officially, 95.5 centimetres fell in Halifax, 40 centimetres more than any previous storm. The wind created snowdrifts several metres high, and snowplows left similar-sized snowbanks for residents to shovel.

Frederictonians kayak through some of the fifty flooded streets in the city's downtown area in New Brunswick. An extraordinary 4-metre winter snowpack rapidly melted during the warm spring rains, pushing the Saint John River more than 2 metres above flood stage.

Water limits access to the Westmoreland Street Bridge in downtown Fredericton during the peak of the flood, cresting at 8.36 metres above average. The floodwater would take a record seventeen days to recede below flood stage.

Photo © The Canadian Press/Andrew Vaughan

Downed power poles litter the streets of St John's after hurricane-force winds from post-tropical storm Leslie were unleashed on September 11, 2012. Wind gusts have been as high as 180 kilometres per hour during storms that frequently ravage eastern Newfoundland. St John's is the windiest city in Canada, with an average annual wind speed of 21.9 kilometres per hour.

About 1,400 commuters were stranded on a GO train when the Don River overflowed during a unique summer thunderstorm. It would take police services hours to rescue all the passengers. Converging winds over Toronto produced the city's greatest one-day rainfall and led to $850 million in damage.

Photo © The Weather Network

Buses and a car on South Common Route 26 struggle through a metre of water in a South Mississauga underpass. The storms would stall commuters across the Greater Toronto Area and lead to power outages for nearly a million customers.

Photo © The Weather Network

The Confederation Bridge has the world's longest span and covers a body of water that regularly freezes. The 12.9-kilometre bridge rises 60 metres above the Northumberland Strait, linking Prince Edward Island to New Brunswick. The fixed link was officially opened on May 31, 1997.

Smoke and flames rise from the Beast near the media assembly area on Highway 63 near Gregoire Lake. More than 3,000 buildings, including 2,400 homes, were razed by the wildfire. The fire cost $9.9 billion, the most expensive disaster in Canadian history.

Seen 20 kilometres from Fort McMurray, on May 6, 2016, a wall of smoke rises from the fire that eventually burned 598,000 hectares of boreal forest in Alberta and Saskatchewan, an area larger than the province of Prince Edward Island. The fire took no lives, although two people died in an auto accident while escaping the flames.

Photo © Chris St. Clair

Hurricanes are becoming stronger and more frequent, repeatedly impacting major urban areas. In August 2005, 80 percent of New Orleans, Louisiana, was submerged under metres of seawater. The death toll was over 1,800, and forty-six offshore oil platforms were damaged or destroyed. Additionally, Hurricane Katrina levelled 5,300 square kilometres of forest in Mississippi.

Photo © The Weather Network

Hurricane Harvey flooded the Texas coast in August 2017, causing $125 billion in flood damage to Houston, which has a metro population of over 7 million. The hurricane was the wettest on record in the United States, producing over 1.5 metres of rain. A warmer atmosphere holds more moisture, leading to greater amounts of rain during storms.

Photo © The Weather Network

During Hurricane Irma the beach had been pushed onto South Fort Lauderdale Beach Boulevard by storm surge (an upwelling of ocean water driven by strong wind and deep low pressure). Earth-moving equipment was employed to clear the roads of sand in the same way snow is plowed in Canada.

Photo © Chris St. Clair

It looked like January in June. Hail blanketed parts of northeastern Calgary during the most expensive hailstorm in Canadian history, with $1.2 billion in damage. Powerful rotating winds slung hail at a 45-degree angle, which caused extensive damage to vehicles and buildings.

Lytton, British Columbia, seen in July 2017. On three consecutive days in late June 2021, Lytton broke the record for the hottest temperature ever in Canada at 49.4 degrees Celsius. It's also the highest temperature recorded anywhere north of latitude 45, and hotter than any temperature ever recorded in South America or Europe.

A massive pyrocumulus cloud rises to the north of Kamloops, British Columbia. The cloud formed from smoke and particulate from a wildfire and rose several kilometres into the atmosphere, attracting minute droplets of moisture or ice. Friction within the cloud created electrical discharges in the form of lightning, which led to the ignition of more fires.

Photo © The Weather Network

Abbotsford, British Columbia, population 150,000, awash in floodwater during the November 2021 flooding of the Fraser Valley. A 2020 flood mitigation plan by Kerr Wood Leidal Associates for the city of Abbotsford suggested that a one-hundred-year flood could overwhelm the systems of dikes, levees, and pumps that drain the lowlands of the former Sumas Lake area.

We spent the next four hours on Fort Lauderdale Beach, sending reports back to The Weather Network and CBC News Network. The condominiums and hotels lining the beach were largely deserted, and the restaurants, cafés, clubs, and stores had been boarded up. There were no cars on Fort Lauderdale Beach Boulevard.

Powerful storms such as a hurricane will impact how ocean currents behave. Changes in atmospheric pressure will influence the sea surface, and the winds associated with these pressure changes will cause water levels to rise in some areas. As the amount of water in the ocean is a near constant, the rising water in one area must be drawn from another area. Think of a bathtub half-filled with water. If you begin circulating the water with your hand, it will rise in one part of the tub and fall in another. The same principle is at work on a grander scale during a hurricane.

Because Irma was still southeast of Florida and hurricanes circulate counterclockwise, the air flow and wind-driven currents were running along the coast, from the north to the southeast, toward the centre of the storm. That movement of water was eroding the sand behind us, and all along the eastern coast of Florida during our four-hour broadcast the beaches had been reduced by a metre. It was a visual clue to the strength of this storm.

We gathered our gear and headed back to the Jeep; it was the only vehicle parked in the public lot adjacent to the beach. While we were loading the truck, a woman approached us. She wore an old blue tracksuit, her hair was messy, and a little girl was holding her hand. The child, who was about seven, was clad in dirty pink pyjamas. They were homeless.

"Do you have any money?" the woman asked.

Mark walked over. "You've got to leave this beach. It's not safe here," he said.

"No, it's safe," she insisted.

"No," Mark reiterated, "there's a hurricane coming. It's too dangerous to be by the ocean."

"When is that? Who told you that?" she said, her eyes darting back and forth between us.

I gave her a twenty-dollar bill and we both stressed again that she needed to leave the area. The woman took the money, the child took her hand, and they walked back to the beach. In the distance, among the palm trees, stood three pup tents. The pair walked toward them, and on arrival several more people stood up and looked our way.

Mark and I locked up the Jeep and walked across the road to Fire Rescue Station 49. We rang the buzzer.

"Hello, how can I help you?" a voice said.

We explained that there were people living rough on the beach.

"We're aware and we'll collect them up when we secure the beach later today."

"What happens to them?" I asked.

"We hand them over to the police" was the response.

Mark looked at me. There was nothing for us to do. We were standing in front of million-dollar condominiums and hotels that sold rooms for five hundred dollars a night. Two different worlds on one beach.

To our southeast, near the island of Great Inagua in the Bahamas, Irma was undergoing a replacement cycle of its eyewall. In major hurricanes, the centre, or eyewall, will contract in size as strong thunderstorms that surround the centre of the storm strengthen and form a circular outer wall of convection. This convective activity moves inward, altering the flow of moisture and momentum of the eye of the storm. The hurricane slightly weakens until the outer ring of storms becomes the dominant feature and the centre eyewall regenerates. Then the storm gathers more strength, or energy. There are several hypotheses as to why this happens, and a clear answer hasn't yet been agreed upon. It is a fact, though, that almost all Category 4 and 5 hurricanes go through this process.

Irma was sporting 270-kilometre-per-hour winds as it moved west toward Cuba. Mark and I were back in the Jeep, looking for a Walmart or Target that was still open. We needed to get some water and food, enough to last us the next few days. We also needed some jerry cans for extra gasoline. The streets of Fort Lauderdale were nearly empty of traffic. Lines of ten to twenty vehicles were at the gas stations.

We bought a couple of five-gallon jerry cans at a Home Depot, then joined the lineup at a Chevron station. While Mark waited to fuel up, I walked next door to a Publix grocery store.

There was a handful of people inside. The shelves were nearly empty. All of the perishable goods had been moved to the front of the store. The prices on milk, meat, fruit, and veg had been slashed. It was better to move it than to have it rot if the power went out.

I filled a cart with a dozen tins of tuna and ham, a box of energy bars, a few boxes of crackers and cookies, a big bunch of bananas for fifteen cents, a dozen Gatorades, and the last three cases of water. I paid up and pushed the grocery cart out the door and into the bright sun and oppressive humidity. I looked for Mark. He was still in line, and it was moving slowly. They'd run out of everything except high test and diesel.

As we waited to get fuel, I called the office. It would be a good idea for us to record some reports on the progress of the evacuation. We spoke with several customers who were in a hurry to fill their cars and get back on the road.

One couple told us that they had thought about "riding out the storm" at home but changed their mind. They had driven north from Miami.

"All the gas stations down there are closed," they said in unison. "It's all closed up there."

At 1:30 p.m. we wheeled onto the interstate to Miami. The long lines of cars that we'd seen heading north when we had arrived some twelve hours earlier had dissipated; now just the occasional vehicle was heading that way and there was no traffic in the southbound lanes. The entire interstate was empty.

It was profoundly odd to drive on an urban expressway alone at two in the afternoon. There was so much evidence of human activity—roads, buildings, warehouses, shopping malls—but there wasn't a soul in sight. Mile after mile we drove not seeing anyone, just exit ramps, overhead signs, and billboards for casinos, lawyers, and cheap insurance.

We pulled the Jeep into the parking garage at the Hampton Inn at Miami Airport. When I had booked our rooms, I had chosen hotels in areas that were most likely to have their power restored first after the hurricane. My logic was that hospitals would get their power back quickly, and so would the radar and other facilities at the airport. Both would be vital services during and after Irma. The four-storey concrete garage was also on my "needs list"; we had to have mobility to cover the storm, and I didn't want the vehicle damaged by a fallen tree or flying debris.

The lobby was crowded. There were an awful lot of seniors milling about the lounge and restaurant.

"There seems to be a lot of older people here," I said to the check-in agent. Of the thirty people in the lounge, only the check-in agent, the valet, the concierge, Mark, and myself appeared to be under seventy.

"Ah," the check-in agent smiled. "Yes, a retirement complex has relocated here for the hurricane. They were flooded the last time we had one." He smiled again.

"So it should be fairly quiet then," I replied.

"We'll see," he said as he pecked at the keyboard and scanned the pass cards.

My conversation was rewarded with two rooms on the eighth floor.

"We want to keep the more elderly guests on the lower floors in case the power goes out. The stairs can be a bit much," the agent told me.

Mark and I retired to our rooms for a quick rest. I spent an hour surfing through the local coverage of Irma and watched The Weather Channel to gather their synopsis of both the storm and the situation in Florida. Access to the barrier islands had been closed, which meant we would not be able to broadcast from Miami Beach or any of the beaches across Biscayne Bay. Route U.S. 1 south to the Florida Keys was prohibited as well. All lanes were going in one direction now—off the Keys. At midnight the road would be completely closed to all traffic. Anyone left behind would have to shelter in place. Irma was still a Category 5 hurricane.

Mark and I reviewed the computer models. The storm was tracking almost due west; its eye was a couple of hundred kilometres north of Holguín, Cuba. A strong ridge of high pressure to the north was impeding Irma's expected turn toward Florida. If this ridge of stable air weakened overnight, Irma would begin veering to the northwest Saturday morning, putting Miami in line for a direct hit on Saturday night.

However, one computer model suggested that the ridge wouldn't weaken until Saturday morning. If that turned out to be the case, Irma would still curve to the northwest, but not until later in the day. The difference of just six hours in the timing meant that the new scenario would place Naples, Fort Myers, or the Tampa Bay area in the crosshairs of a direct impact. This was a critical point; many residents from Miami, Dade, and Broward counties had evacuated to those cities out of habit. Historically, a hurricane that impacts Florida usually arrives from the Atlantic, travelling in a northwesterly direction, or from the Gulf of Mexico, in a northeasterly direction.

A vexing set of meteorological conditions was playing out, making it increasingly difficult to determine exactly where Irma would go and when the storm would arrive there. Regardless of the exact landfall location, however, all of Florida would be impacted with flooding rain and damaging wind. The outer bands of this sprawling storm would bring thundershowers by Saturday morning.

The landfall was what interested Mark the most. Landfall is when the eye of the hurricane arrives on shore. This arrival is preceded by powerful wind, rain, and storm surge. As a storm chaser he wanted to be in the right location to witness the eye of the storm. It is rare to find yourself in that place, but Mark had done it several times before and was hoping to relive the experience with Irma.

The eye of a hurricane extends from the surface upward. It is ringed with the fiercest wind and heaviest rain. Inside this ring lies a calm; the rain ends, the wind abates, and as the 30- to 60-kilometre-wide eye passes, one can look straight up and see blue sky. The passage of the eye could take anywhere from twenty minutes to an hour, depending on the speed of the hurricane. Eventually, the eyewall will overtake you and replace the calm with a maelstrom of wind and rain.

We called our meteorology team in Oakville to discuss the hurricane. They said a consensus was forming among forecasters that Irma would make landfall by Saturday night, somewhere between Miami and Naples. It was as tight a fix as they could make with the available data. The distance between the two cities is 160 kilometres. A slight shift in the storm's path could move it 60 kilometres in either direction. We'd review the forecasts again in twelve hours. Perhaps the world's most powerful weather computers would offer more clarity in the morning.

We headed out of the hotel and back to the Jeep to do our live reports. The sun was shining, but the streets were deserted. We drove east on NW 7th Street. The shops and restaurants in Little Havana were boarded up; windows that weren't covered with plywood had been crisscrossed with duct tape. Gas station pumps were tightly bound with a thick wrapping of cellophane. We saw no one as we drove the

10 kilometres to Bayfront Park across from the Miami cruise ship terminal.

"Imagine being in downtown Vancouver on a Friday night, standing at Canada Place, and the city is basically empty of people. That's what it's like here in Miami tonight," I said during our live report.

Mark was standing beside me. "It's not just here," he said. "It's like this all the way up the coast to Melbourne. Irma is dangerous for a couple of reasons: yes, the wind, but also the rain could be a huge problem for the Gulf Coast. Just two weeks ago, Tropical Depression Ten flooded the area from Naples to Tampa—during this storm, a half-year's worth of rain had fallen in just two days. That water still hasn't been absorbed, so any more rain will immediately lead to overland floods."

We wrapped up our broadcast at around ten that night.

"Let's go back to Bayfront Park tomorrow," I said on our drive back to the hotel. "The difference in the weather from tonight to the morning will be interesting."

"Maybe even the first thunderstorms will roll in on the outer bands by eight," he said.

When we got to the hotel, we gathered up all our equipment and headed to our rooms. The lobby was empty except for the desk clerk.

"No parties tonight?" I said.

He looked up and smiled at us. "No, very quiet."

"Meet you in the lobby at five. Get some rest" were my parting words to Mark.

As we closed the doors to our rooms, Irma was making landfall in Cuba near Cayo Romano, which is just to the east of Cayo Coco, a popular sunshine holiday destination for

Canadians. The wind was measured at 270 kilometres per hour, making Irma only the second Category 5 hurricane to strike the island since 1924.

My phone woke me at 4 a.m. and I turned on the TV. The latest analysis of Hurricane Irma was being discussed by a panel of experts from the National Hurricane Center in Miami.

Irma was still moving to the west. Its forward speed had slowed as it was now interacting with the Cuban coastline, but there was still a good opportunity for the storm to arc northward toward Miami. However, the experts said, there was an equally good chance that Irma would push toward the Gulf Coast and come on shore near Naples. Either way, Florida was going to receive the same battering being dished out in Cuba.

Mark and I met at 5 a.m. and loaded the Jeep with our equipment. As we pulled out of the parking garage, a light rain began to fall. It was as humid as it had been the previous day.

We drove east on 1st Street, through the canyons of office towers in the downtown core. Flashing red traffic signals reflected off the wet pavement. We turned onto Biscayne Boulevard. Police barricades had been erected along the entryway to the parking lot at Bayfront Park.

Mark stopped the Jeep. The headlights pierced through the drizzle. We got out and began looking for someone to talk to about accessing the park.

There were no police cars or officers in sight. We looked at each other and I said, "Let's go for it." I pulled a sawhorse barricade away and Mark steered the Jeep through the opening. I then closed the barricade and climbed back into the truck.

We followed a service road to the boardwalk that ringed this part of the harbour, then drove down the boardwalk to one of the several jetties that, on nicer days, offered tourists rides on Cigarette boats around Biscayne Bay. We parked the Jeep and began setting up the camera and Dejero unit.

I made a phone call to the Oakville office and let them know that we were set up and ready to go. We would have two background scenes for our live segments: one was the lights of the Miami skyline; the other would be the cruise ship terminal. The ships had left to ride out the storm at sea, so the location would look much better once the sun rose in an hour or so.

Just after 7 a.m., while we were live on The Weather Network, several black GMC Suburbans pulled up to the jetty beside our location. Mark and I continued our live segment while a satellite truck backed onto the jetty behind us. The Weather Channel had arrived to do their live coverage.

It was a very interesting scene. We, at The Weather Network, were really lean. It was just the two of us, with the suitcase-sized Dejero unit, a camera on a tripod, and the Jeep. On the next jetty were a satellite truck and its operators, three cameras and their operators, two producers, a microphone boom operator, a complete lighting set, a teleprompter and its operator, and Jim Cantore, the host of The Weather Channel.

While both networks did simultaneous live broadcasts from Biscayne Bay, a dozen wild roosters strutted behind us and crowed out "cock-a-doodle-doo." Thundershowers passed over us several times, with heavy downpours and gusty winds. The water was getting choppy.

More media arrived. NBC had set up its Miami coverage beside The Weather Channel, and three local stations—Fox, ABC, and CBS affiliates—had also arrived with their satellite trucks.

Irma, the focus of all this media attention, was 400 kilometres to the southeast. The eye of the hurricane was meandering along the Cuban coast, sometimes making landfall and drifting inland, then slipping back over the sea again. This interaction with land was beginning to weaken the storm. The winds had decreased to less than 240 kilometres per hour. The once-visible eye was becoming obscured by cloud.

Sitting in the Jeep, with rain pelting the windshield, we held a conference call with our meteorology team and producers. The latest computer models were in, and a storm hunter Orion aircraft had recently conducted a reconnaissance flight.

Storm hunter aircraft are used for hurricane research. They are operated by the National Oceanic and Atmospheric Administration, or NOAA. Experienced crews of scientists and aviators fly directly into the eyes of hurricanes, taking measurements of the air and water. The data gathered by these flights is critical to understanding and forecasting the future behaviour of these massive storms.

The latest data from the Orion flight, when combined with the new computer models, strongly suggested that Irma would begin turning to the north late on Saturday, which would take the eye of the hurricane west of Miami, over the Florida Keys, with landfall expected early on Sunday, somewhere between Naples and Sarasota on the Gulf Coast.

We quickly recorded three reports about the latest forecast for the storm and then loaded up the Jeep with our gear. After a quick stop at the hotel to gather the rest of our belongings, we headed west on the I-75 toward Naples.

Our meteorology team called us as we were nearing Naples. After much review of the forecast, they felt that the Tampa–St. Petersburg area was of high probability for a landfall on Sunday. We continued north on the I-75, past Naples, to Fort Myers. Mark was working his network of storm chasers, who were now cruising up and down the coast, looking for a good location to catch the arrival of Irma.

The Jeep was down to less than a quarter of a tank of gas. I decided to pull off the interstate at the Fort Myers airport exit to find some fuel. The service stations were shuttered, their gas pumps wrapped in cellophane to protect them from the elements. We drove to the terminal. It too was closed and locked up.

Mark called his storm-hunting buddies. They were in Fort Myers and gave us the locations of two gas stations that were still selling fuel. We drove down Colonial Boulevard to one of them, where we joined a long queue.

A Toyota SUV pulled into the parking lot beside us. A weather station was mounted on a roof rack, two large jerry cans were attached to the rear, and an array of CB radio antennas adorned both sides of the truck. Mark's three storm-hunting friends stepped out of the truck and greeted him with hugs and slaps on the back. I was introduced to them: Scott McPartland and Dave Lewison had flown in from New York, and Tim Millar was a freight pilot who lived in South Florida and regularly gathered storm information for the National Hurricane Center.

Over the past dozen years this group had chased numerous hurricanes and tornado outbreaks throughout the United States. They were all experienced storm chasers who, in addition to enjoying the exhilaration of extreme weather, contributed to the meteorological and scientific study of those costly and dangerous events. The three had found an open pizza place and wanted us to have "a couple of pies and talk about Irma."

We got back into our trucks and drove several blocks to an Italian restaurant located between a laundromat and a shuttered 7-Eleven in a seedy strip mall. Ours were the only vehicles in the lot.

We took a booth in the corner of the restaurant. The discussion centred on the best location for the hurricane to pass, so they could stand in the middle of the eye and get a spectacular view of the clear sky directly above. It was still too early to determine where that would be right now, but they had scouted out several parking garages up and down the coast.

"Do you have a gun?" one of them asked.

"No, we're from Canada, remember?" Mark replied.

"Do you want to borrow one?"

I asked why we'd need one. The answer was straightforward: "To protect yourself. You never know what might happen in all of this."

Mark and I laughed it off. We paid the bill and stepped outside into a blanket of humidity. The others planned on heading to Marco Island near Naples; we would drive north toward the Tampa Bay area.

In the car, I dialled into an all-news station providing full Irma coverage; in the Tampa Bay area and along the Gulf Coast, access to all the barrier islands would be closed completely that afternoon.

"Let's pull into Sarasota and make that our base," I said.

We pulled up to our hotel, chosen because it was brand-new

and built according to the very latest building codes. Those codes had been updated to consider the strength and frequency of hurricanes in the region. It was an excellent example of how many businesses and industries were being proactive about the impacts of climate change.

Children were playing tag in the lobby. The hotel was full of young families who'd come here to escape the hurricane. The clerk explained that the hotel generator would provide electricity in the common areas if the power went down.

Television monitors in the atrium played CNN, The Weather Channel, and the Cartoon Network. It was three in the afternoon.

We had an opportunity to talk with some of the guests. Half had migrated here from Miami; the rest were locals, including several Canadian families who owned homes in the area. The locals had sought refuge at the hotel because they feared their homes would flood or be structurally damaged by the winds.

The news that we relayed that evening was not good. On Cayo Romano in Cuba, storm surge had driven 8-metre waves on shore. The ocean water flooded inland for nearly 2 kilometres. Some weather stations were reporting over 600 millimetres of rain, which is 2 feet in one day. An anemometer in Ciego de Ávila had recorded wind velocity at 256 kilometres per hour. That relentless wind had torn the roofs from over 23,000 homes; another 14,000 homes were destroyed.

As we wrapped up our broadcast from the base of the now closed John Ringling Causeway, Irma began its turn northward toward the continental United States. The water between Cuba and Florida was warm. The hurricane gathered energy from the sea and intensified in strength to Category 4; winds were up to 250 kilometres per hour.

Landfall would come the next morning.

We woke before dawn and drove back into downtown Sarasota at five-thirty. A steady rain was being blown by a strong easterly wind. As we slowed at intersections, the flashing red lights overhead swayed precariously and trash bins rolled up the street. Palm fronds lay scattered across the ground.

Overnight, Irma had crossed the Strait of Florida and, as our forecasters had suggested, was now a Category 4 hurricane. The eye was well formed and clearly visible on the radars located in Miami and Tampa.

Irma would make its first landfall in the Florida Keys just after sunrise.

We manoeuvred the Jeep around a fallen tree at the marina on Bayfront Drive and parked it next to O'Leary's Tiki Bar & Grill. The windows and doors of the restaurant were secured with plywood and two-by-fours. Picnic tables and umbrellas were piled along the wall.

Wind made opening the truck's doors difficult, and the rain found its way through our rain jackets. We were soaked by the time we'd set up our gear under an awning at the bar. I snapped a photo of Mark standing next to the sign and posted it on Twitter with the caption, "Is it too early for a beer?"

My earpiece came to life with the voice of our producer at The Weather Network. "You're on at 6:03 a.m. We can see you on the monitor," she said.

I gave them a thumbs-up.

Irma provided our audience with increasingly good visuals that morning. Waves were crashing over the seawall behind us, and boats slammed into each other in the marina. One sank as the relentless waves splashed over its gunnels, filling it with enough seawater to force it under.

At 7 a.m., with winds of 215 kilometres per hour, Irma crossed Cudjoe Key, about 15 kilometres east of Key West. It was the first landfall in the United States, and almost every structure on the island was obliterated.

Social media quickly filled with video of the hurricane's approach to the Keys—until the cellphone towers fell to the powerful gusts.

The winds from Irma were the strongest to strike Florida since Hurricane Charley in 2004, and its barometric pressure was the lowest recorded in the state since Hurricane Andrew struck the south side of Miami in 1992.

During our broadcasts from the marina, we were able to witness an amazing phenomenon: over the course of an hour, a reverse storm surge took place along the Gulf Coast. The water level of the Gulf of Mexico began to decrease as circulation around the eye of the storm drew the atmosphere and the ocean waters toward its centre. It looked very similar to the low tide that occurs in the Bay of Fundy and Minas Basin in Canada. Over ninety minutes, the water drained from the harbour. Sailboats, cabin cruisers, and the buoys that marked the marina were now sitting on the sea floor.

I did a live report, walking across the bay in water that barely went up to my knees. I stopped at a sailboat that lay stranded, tipped on its side and grounded on the bottom of the basin. I emphasized how many trillions of litres of water had been displaced—more than the equivalent of all the water in Lake Erie. It was astounding.

Word quickly spread to those who lived nearby. Soon hundreds of people were walking in the very shallow water at Bayfront Park, posing for selfies while marvelling at this natural phenomenon.

Meanwhile, the opposite effect was happening on the east coast of the state. The 2 metres of water that had vanished in front of us was showing up an hour later in the form of a 2-metre-high surge of water in Miami and communities as far north as Jacksonville. The water didn't come as a tidal wave; instead it rose gradually over the course of several hours. The powerful winds exacerbated the impact of the rising ocean, driving seawater up rivers and canals and flooding parts of downtown Miami.

In six hours Hurricane Irma crossed the 150 kilometres from the Keys to Marco Island, near Naples. The storm was moving faster now, the acceleration triggered by an infusion of dry and slightly cooler air from the continent. That air mass was also beginning to weaken Irma.

When the eye crossed over Marco Island at 1:30 p.m., the wind was measured at 185 kilometres per hour. Irma was a Category 3 hurricane now, and would quickly abate as it moved northward over the course of the afternoon.

Cellphone and power transmission towers began falling throughout southwestern Florida; by mid-afternoon it was becoming difficult to get information from the Naples area, where 75 percent of the towers were not functioning. Out on the Florida Keys, 89 of the 108 cell towers were offline.

Wind-driven rain buffeted Mark and me as we broadcast

live from Sarasota. Behind us the streets were filling with what storm hunters refer to as "salad"—an accumulation of tree limbs, leaves, and palm fronds—which impeded drainage and led to flooding on many roads.

Two weeks earlier this same area had seen flooding as a result of Tropical Depression Ten. Some locations had measured over 400 millimetres of rain. As a result, when rain from Irma poured from the sky and pooled on the streets, the ground was still saturated. Any low-lying surface quickly became a pond or small lake.

As darkness fell, the power in Sarasota began failing. A lot of cellphone traffic was being fed through a decreasing number of transmission towers, and data service became increasingly slow.

By 10 p.m., we were out of the live broadcast game. It had become impossible to transmit images from the Dejero. Even our iPhones signalled no service.

We packed up our equipment and began the drive back to the hotel. The power was out, and the Jeep's headlights pierced the darkness and heavy rain. Wind-blown debris banged off the roof and sides of the truck. If it made a clanging sound, it was likely a garbage can or piece of aluminum siding; a dull thump usually indicated a palm frond or tree branch.

We passed dozens of trees that had been felled by the wind. I slowed the Jeep to navigate around branches and portable signs that littered the streets. We occasionally came across the flashing orange lights of a power crew removing downed electrical lines from the roadway.

It took forty minutes to make the drive back to the hotel.

The ragged eye of a weakening Irma was 60 kilometres

southeast of our hotel. Most of the newly planted palm trees that ringed the parking lot had been blown down. Some lay over the crumpled roofs of cars; others had missed their mark and fallen on the driveway. The wind made a steady roar, and the light standards over the parking lot swayed; gusts blew the rain sideways.

The hurricane was still delivering 170-kilometre-per-hour gales. But the constant interaction with terrain was sapping Irma of strength. The topography and mechanical interference of buildings were beginning to dampen the winds.

Land could not offer Irma what the ocean had—the heat energy to nourish its growth. Now over a large landmass, the hurricane's powerful circulation was pulling dry and cold air into the heart of the storm. In the next twenty-four hours Irma would devolve from being the most powerful storm on Earth to being just another large low-pressure centre bringing heavy rain and gusts to Georgia and Alabama.

I fell asleep listening to the wind rumbling and rain pelting the window of my hotel room.

At 4:30 a.m., "Holiday Inn" by Elton John began playing on my phone. It's the song I use on my alarm when I travel. I listened to the whole song before I picked up the phone, which showed one bar of service.

"Great," I said out loud as I turned on the light. No luck, the power was still out, which meant no coffee, no shave, and no warm shower.

I called The Weather Network and spoke with the producer. "Yep, we're all good. The power is out, but I've got one bar of cell service. Maybe we can manage to get on air with the Dejero." Together we hashed out a plan to join the morning show at six. I called Mark with the update, then packed my suitcase and headed to the lobby.

Generators were supplying light to the halls, lobby, and self-serve kitchen. I made breakfast and had coffee while I waited for Mark. At 5:45, I went for a walk. It was still dark, but the rain had stopped. The air felt less humid and the winds had calmed. I walked over to a palm tree that had fallen onto a car. The tree was about 15 metres long; the crushed car was a new BMW 7 series. We would do our live reports from here.

As the sun began to rise and tint the eastern sky orange, we began our report.

"On average, 75 percent of all the cellphone towers from Sarasota to Naples are either down or without power; 6.7 million customers are without electricity; thousands of crews are now working to get the downed lines cleaned up and new lines strung," I said while video of the destruction of the Florida Keys played on screen.

The outage was one of the largest disaster-related power losses in American history.

Mark and I recorded several reports for the network, loaded our gear into the back of the Jeep, and emptied both jerry cans into the tank. Then we set off southbound toward Naples. The fuel gauge registered barely half a tank.

Water filled the centre median and rose to the pavement on both sides of the interstate. At times it completely breeched the shoulders and the highway became a shallow lake. We slowed as we drove through the water, which was about 20 centimetres deep on the road and closer to 2 or more metres deep if you stepped off the shoulder.

I looked at Mark and said, "Alligators." They would have been displaced by the flooding, and it would be important for us to stay alert when we got out of the Jeep.

Convoys of power trucks overtook us, coming in groups of ten. We saw well over a hundred on the drive south. The sun beat through the window, and the sky was clear and blue as we passed Port Charlotte and Fort Myers.

We rumbled down the highway looking at damaged buildings, their roofs torn away and resting against a neighbour's home, walls collapsed with insulation strung like cotton candy in whatever trees were left standing. After an hour, it began to look normal; we were becoming accustomed to the vulgarity of destruction.

Other than the convoys of power trucks, there were very few vehicles on the highway. A few state troopers passed us in their cruisers across the pond that separated the lanes of the interstate. A couple of ambulances drove north, lights flashing.

We eased the truck onto the exit ramp and turned west toward downtown Naples. On our right was a trailer park. It appeared that the mobile homes had suffered severe damage from the hurricane. Strips of aluminum hung from the low-slung date palms; an entire trailer lay open—no roof or walls, just someone's belongings spread across the property, their life on display for all to see.

Slowly, we continued west.

Our progress was limited nearly immediately. The power was out, and a police car was parked with its lights flashing in the middle of the first intersection that we arrived at. The officer on duty held up his hand and signalled for us to pull over. As we complied, he walked up to my window. His hand moved to the holster on his hip, and he unbuttoned its clasp. He was six feet tall, had dark hair, and wore wraparound sunglasses and a bulletproof vest.

"Slowly, take out your vehicle registration and show me your driver's licence," he said.

I handed over my Ontario licence while Mark found the registration and rental papers for the Jeep.

As he looked through the papers I asked him, "What's it like in town? We're news media from Canada and would like to look around and file a report."

His radio came to life while I was speaking, and he unclipped the microphone from his vest and responded to the call.

A convoy of yellow power company trucks rolled up behind us. The cop waved them through the intersection and then turned back to us.

"You can't go any farther," he said. "It's a state of emergency. Only residents who live here can have access. It's too dangerous with the downed power lines and looting that's been going on. You can help us by getting back on the highway and staying out of town."

He handed me my licence and our papers, and pointed back toward the interstate.

I turned the Jeep around.

"Let's check out that mobile home park," I said. "There might be some good stories about what it was like there when the storm hit, and we'll get some good video."

We pulled the truck into the trailer park and drew to a stop about 10 metres down the lane. The road was flooded and downed trees prevented any further progress in the Jeep.

As we stepped out of the car, a young boy on a bike rode past us and around a sofa that blocked the road. Two more children followed him on their bikes, and they turned into a driveway farther down the laneway.

We unpacked the camera and followed them into the front yard of a double-wide mobile home. The entire side wall was missing, and the roof sagged down on the battered front porch. Propane tanks and kitchen furniture—a table and two chairs—lay in a puddle by the front steps. All of the windows were broken and the curtains fluttered in the warm breeze.

It was humid, and the air carried the unmistakable aroma of sewage. Puddles of water spread from beneath the trailer to a larger pond that was forming in the ditch beside the road.

Mark walked toward the trailer and called out, "Hello, is anyone there?"

The kids we'd just seen and a woman who must have been their mother appeared from behind the trailer. They all looked dishevelled and tired. The woman looked worried and despondent.

"How are you doing? Are you okay?" Mark asked.

"Well, we're alive," she said.

We explained that we were from a television station in Canada and would like to interview her.

She perked up. "Will I be on the TV news here?"

"No, it's for Canada," I told her.

A few of her neighbours began to emerge from their trailers, and several more children came to see what was going on. Insulation fluttered from the broken-down walls of their mobile homes, and plastic lawn furniture lay on the ground and floated in ditches.

I hoisted the Sony camera onto my shoulder while Mark grabbed the microphone and stood next to the woman. She was in her thirties, had peroxide blonde hair, and wore a Van Halen T-shirt, track pants, and plastic flip-flops.

I adjusted the camera to frame the battered mobile home behind them. The kids began playing peek-a-boo out of the broken windows. *This will look good*, I thought, while counting Mark in: "Two, one, and tape rolling."

"Our truck wouldn't make it to Valdosta, up in Georgia. It's where I'm from," she told us. "So we just stayed here and rode it out."

Her name was Cindy, and she worked at the Walmart Supercenter, just down the road. Four of the kids in the trailer were hers. Her husband was working.

"He's a janitor at the La Quinta motel downtown," she told us.

"It's not right, what's going on here," she continued. "All those power trucks go right downtown first. They go right to the gated places. They don't stop here. It's not right. Why do they get power first and I have to wait for two or three days? Why them and not me and my family?" There were tears in her eyes; she was angry. "Why won't someone help me? I need help, please."

The interview had drawn out more neighbours. A half-dozen adults stood scattered along the laneway.

"You need to tell people what's happening here," a big, bearded man told me. He was wearing a well-worn ball cap, jeans, and a stained, sleeveless T-shirt. "This ain't right," he said. "No one's been here, not even the police. Are we just supposed to fend for ourselves? Where's the government to help us? Do you see them here helping us?"

Their collective anger grew, feeding from one person to the next and escalating each time someone began to vocalize their frustration. I dropped the camera off my shoulder and gripped its handle with my left hand. Mark walked toward me. We both felt the shift in tension.

"Listen, folks, thanks a million for talking with us. We really appreciate it," I said. "We'll do what we can to get you some help right away."

"Do you have any water?" the bearded man asked.

Mark replied, "Yeah, we do. It's in our truck. We've got a couple of cases. You can have them."

The bearded man and two of his buddies walked back to the Jeep with us. Picking our way around sheets of vinyl siding and downed trees, the five of us walked single file around the stagnant pond on the road, the septic smell wafting up our noses as we passed.

"You've gotta watch for gators now," one of the men said as we walked down the shoulder of the road.

The sun shone brightly. It was midday, warm, and quiet as I opened the rear door of the Jeep. One new case of water and another dozen bottles sat in the back of the truck; beside them, a box with our tinned tuna, cookies, and power bars.

The bearded man stepped past me, his pungent body odour mixing with the aroma of ruptured septic lines. He reached into the back of the Jeep.

"We'll just take all of this," he said as the three of them gathered up the water and food.

One of his buddies spoke for the first time. The man was in his thirties and tall. His thin arms were covered in tattoos, mostly pentagrams and snakes. He looked like he'd seen more than his share of hard nights and bar fights.

"You probably want to give us some money now too. That'd help us out a lot," he said.

They all looked serious about the request—serious but not necessarily menacing.

I stepped back from the three of them and said, "We don't carry cash. In our work it's just not safe to do that. Sorry. You know, I hope the water and food can help a bit, and tell Cindy thanks for the talk."

I pulled the back door shut and said, "Let's go, Mark," then turned and walked to the driver's-side door, opened it, stepped inside, and started the truck in one fluid motion.

Mark was pulling his door closed as I dropped the Jeep into drive and aced a three-point turn in under eight seconds. We were on our way again.

I looked at Mark and said, "There is no way that having a gun would have helped how that all ended."

"No shit," he replied.

We drove back to the intersection to tell the cop about the people in the trailer park and how they felt they needed some assistance. He asked us if anyone was injured.

"We don't think so," I said.

The officer said he'd have someone check in on them when he had a chance.

It was time to hit the interstate again, eastward. We'd used half the fuel. The tank now registered a quarter full, and we still had no cellphone service. We tuned into the local news radio stations, which confirmed the obvious: the power was out for all of southwestern Florida. Electricity was gradually being restored, but it would take time. Things were marginally better in the Fort Lauderdale area, which had incurred less damage. That would be our next destination.

Both of us were doing fuel mileage calculations in our head as we drove straight east into the Everglades. I knew there was gas in 100 kilometres.

Forty-five minutes later, we approached the Miccosukee Service Plaza. A roadside sign displayed the price for a gallon of gas and then flashed in bright green lights the price for diesel. They had power!

I pulled our truck off the interstate and joined the line of vehicles waiting to fuel up. There were about twenty cars ahead of us. We pulled out our phones; two bars were displaying. There was cell service here, too.

I called the office to let them know we had service again and were going to file a report and feed them the interview from the mobile home park. While Mark gassed up the Jeep, I set up the Dejero unit and dialled in our Oakville control room. Then I linked the camera to the unit and began feeding the scene from an hour ago.

We decided to do the Canadian thing and take just a half-tank of gas. The lineup was growing longer, and the city was just over 100 kilometres ahead of us. Once we'd pumped, we moved the Jeep into the parking lot and pulled out our equipment.

I set up the camera on its tripod, then both of us stood six paces in front of the Sony. The line of cars and the gas station were framed behind us.

"On you in three, two, one," John, the producer, said in my earpiece.

"Hey, Chris, welcome to the middle of the Everglades, the

only place we could find electricity, cell service, and gasoline," I said to Chris Mei in our studio back in Canada.

Mark quickly added, "Irma passed 50 kilometres west of this spot just one day ago. It's simply amazing there's not more damage here. The big sign and the roof over the pumps have blown away, but that's about it."

We talked with Chris about the scenes of destruction as video played from the Florida Keys and Marco Island. Irma was directly responsible for taking seven lives in Florida; another seventy-seven deaths in the state were indirectly tied to the hurricane. Hundreds had been injured.

After our report we packed the gear back into the truck and rolled onto the I-75 headed east, toward Fort Lauderdale. There were more cars on the highway now, going back to the city. I began checking my phone apps for a hotel. I found two rooms at the Holiday Inn near Fort Lauderdale Beach, a block from where we had begun reporting on Friday morning.

When we checked into the hotel, Mark heard from the storm chasers. They'd been in the eye, and he was going to meet them to compare notes.

I needed to take some time to collect my thoughts and mull over what we had witnessed the past several days. I walked back to the intersection where we had begun our broadcasts on Friday. A single police car with flashing lights was parked in the centre of the road. The officer inside was looking at his phone and didn't stop me as I crossed the road and walked to the bridge connecting the mainland to the island.

A cruise ship was slowly edging toward the terminal to the

south of my vantage point on the sidewalk at the apex of the span. I watched as it returned from riding out Irma at sea. Tomorrow or the next day, it would load tourists bound for another cruise in the Caribbean.

Hurricane Irma's impact on infrastructure across the state would eliminate over 100,000 jobs from the tourism sector in the coming weeks. Refurbishing damaged buildings would continue for at least a year.

I walked off the bridge and along Seabreeze Avenue; the roadway was littered with the familiar salad of tree limbs and palm fronds. As I negotiated my way around the debris, small groups of people were surveying their property for damage and removing the plywood sheets that had shielded the expensive homes.

The structural damage was minimal compared to the trailer park we had visited in Naples. Chainsaws buzzed as trees were disassembled to manageable sizes and then stacked at the ends of front lawns. A Rolls-Royce Silver Shadow parked in a driveway had been crushed beneath a large palm tree.

On South Fort Lauderdale Beach Boulevard, bulldozers and construction graders were plowing sand off the street and toward the beach the same way we use snowplows to clear Lake Shore Boulevard in Toronto. Instead of snowbanks, massive piles of sand lined the sidewalk, right up to the front doors of the hotels, condos, and restaurants that lined the street.

I stopped and spoke with a city foreman who was supervising the project. He told me the power had gone out but was slowly being restored. All of Fort Lauderdale should be back on the grid in two or three days. "There are some places where more trees had come down, and it'll take a few more days to restore power there," he told me.

A coroner's report would later reveal that at least eight seniors in a Hollywood, Florida, care home had died due to stifling heat and lack of electricity. The long-term secondary death toll would continue to grow in Florida and throughout the Caribbean.

The beach was deserted as I walked back to the hotel. There was no sign of the homeless woman and her friends, no evidence of their tents or belongings.

I stood in the sand at the water's edge looking at the Atlantic. The waves sparkled in the setting sun as they rolled relentlessly onto Fort Lauderdale Beach. Another cruise ship was nosing toward the inlet and its waiting berth at the terminal. After Irma, people were doing what they always do—getting on with life.

The 2017 hurricane season would go on to become the most expensive to date, with monetary losses totalling nearly US$295 billion. The cost in human life was staggering. At least 3,364 people died in the seventeen named storms, ten hurricanes, and six major hurricanes that year.

Hurricane Maria, which formed just a week after Irma dissipated, was the strongest of them all. Maria took over three thousand lives and caused US$91 billion in damage alone. Dominica, Guadeloupe, the Virgin Islands, and Puerto Rico bore the brunt of its wrath. In Puerto Rico a humanitarian crisis developed when the island's power grid was destroyed, leaving all 3.5 million residents without power. More than a third of the population would still be without power six months later. Eighty percent of all agriculture was destroyed, and 100,000 people left the island permanently.

There is no question that our warming climate is producing stronger hurricanes and typhoons. In my twenty-five years of covering these types of storms, each season always had its "most devastating storm." But over the past several years, the disturbing reality is that hurricane seasons are now regularly generating *several* catastrophic storms.

Lives, economic damage, and destruction of both property and natural habitats is the price we are now paying for the climate crisis.

CHAPTER 11

Hailstorm Alley

Calgary, Alberta, 2020

The spring of 2020 was wetter than usual in southern Alberta. Grey skies mirrored the prevailing mood as COVID-19 led to shutdowns, uncertainty, stress, and a sense of foreboding for many. May offered sunnier skies and warmer temperatures, but by mid-month powerful storms were forming due to the high moisture content in the soil.

Our reporter in Alberta, Kyle Brittain, is an experienced storm chaser, having spent thousands of hours studying and documenting severe thunderstorms and tornadoes in the province. Kyle also spent years as a forest firefighter, jumping and rappelling from aircraft into raging fires across western Canada. His understanding of the subtle nuances that drive storms in Alberta is invaluable.

June, July, and August are the peak months for thunderstorms and hail in southern Alberta. They are the warmest months of the year, but they are also the months when conditions are ideal for evapotranspiration to occur.

Evapotranspiration is the addition of moisture to the atmosphere from plants and surface water. In the arid farmland of Alberta, crops and irrigation provide the moisture to fuel storms. The additional moisture from heavy spring rains ushered in an early start to the storm season in 2020.

Thunderstorms form when moist warm air is propelled upward into the atmosphere by a catalyst such as convection or a cold front. If the atmospheric temperature decreases rapidly with altitude, a thunderstorm may develop.

The clouds that develop as a storm begins are literally tons of water droplets. However, updrafts within the cloud structure prevent the drops from falling to the ground. As the clouds move higher into the atmosphere, the water droplets at the top of the clouds, where the air is cold, turn to ice. The constant motion and collision of the ice and water particles create an electrical charge inside the cloud structure. The imbalance between electrical charges within a cloud, between two clouds, or between the surface and a cloud will create a discharge of electricity. That discharge is lightning. Thunder is the sound of air exploding as it is superheated to nearly 30,000 degrees Celsius by the electrical discharge. A time comes when the updrafts in our storm cloud can no longer support the weight of the ice and water droplets, and gravity will take over. The particles begin falling toward the surface, the ice melting into rain as it descends through the warmer air near the ground.

In some conditions the storm will continue to grow, perhaps aided by wind shifts or temperature changes. Ice that is falling from cloud tops melts slightly but is caught in a stronger updraft and sent back aloft to refreeze. Additional particles of ice are captured within turbulent eddies near the top of the cloud structure, where they attract water droplets, becoming ever larger pieces of ice. This is how scientists believe hail forms and grows, from the size of a pea to as big as a baseball or larger. The largest known hailstone in Canada weighed 290 grams—over half a pound—and fell in Cedoux, Saskatchewan, on August 27, 1977.

Calgary is situated in Canada's "Hailstorm Alley," an area of Alberta that extends from Lethbridge to Calgary, Red Deer, and Edmonton in the north. This region sees more hail than anywhere else in Canada.

Hail has a proclivity to develop here because of the location. Calgary is 1,045 metres above sea level, which allows for a steeper temperature profile in the atmosphere. High-level winds that flow over the Rocky Mountains, as well as dry, cool winds from the north, often converge in this area. Daytime heating and moisture at the surface are the final ingredients to brew up impressive thunderstorms.

On July 16, 1996, a series of thunderstorms formed along a frontal boundary, a line of cool, dry air that sagged southward over Calgary. Shortly after 6 p.m., hail began to pummel the city. It lasted for several hours, and some stones were as big as softballs. City streets were flooded when the drainage system clogged and water levels quickly rose to nearly a metre. The sheer size of the hailstones tore down power lines and briefly disabled 911 service. Insurance claims made this storm Canada's most expensive natural disaster, with over $300 million in damage.

This type of storm is a regular natural occurrence in this area. What has changed is our human footprint.

In 1996, 800,000 people lived in Calgary. Today it is a city of 1.37 million people living in an area of just over 800 square kilometres. When you add in its exurbs, the area grows to nearly 5,000 square kilometres, another quarter of a million people, and a lot of new homes.

On Friday, June 12, I was working at the office when a new video from Kyle played on my monitor. Thunderstorms were developing, and much like the day before, they were forecast to produce hail and possibly tornadoes.

When a storm is imminent or occurring, warnings are automatically issued by Environment and Climate Change Canada. The maps we were using on television that afternoon automatically changed from "Watch," which means conditions are favourable, to "Warning."

The Weather Network airs five different feeds simultaneously: a national broadcast and four regionally focused broadcasts. When the severe storm warnings were issued for the Calgary area, the national feed and the regional feed for Alberta immediately switched to live coverage; the same broadcast was also uploaded to our phone app for Alberta users.

The storm produced hail the size of golf balls, and a few funnel clouds formed in the sky above the Calgary area. The spiralling clouds hung beneath a dark overcast that moved from west to east over the city.

Fortunately, the twisting funnels stayed high in the sky. The elongated shelf cloud passed over Friday-afternoon commuters, drenching vehicles and causing localized flooding in the water-prone underpasses of the highway system that rings Calgary. Hail fell on some neighbourhoods, but damage was minor compared to past storms. It was a typical severe thunderstorm in Calgary, the second in as many days.

By early evening the cluster of storms had drifted east,

away from the city. As the sun set, eliminating the source of energy to fuel the storms, the air calmed. It was a warm and pleasant night.

Early on Saturday morning, producer Leanne Ferrante and I reviewed the weather maps for southern Alberta: wind and humidity forecasts, radar prediction models, and vorticity outlook models suggest where tornadoes may form. We were hoping we'd be able to get a good idea of exactly when and where the strongest storms would break out later that afternoon.

Isolating the exact time and location of a severe weather event that may be just 10 square kilometres in size in a 5,000-square-kilometre area is difficult work. Computer modelling helps narrow the location, but precisely forecasting thunderstorms and their severity is like playing three-dimensional chess. Local experience and knowledge of an area is just as critical.

We called Kyle to talk through the forecasts for that afternoon and evening. The computer models indicated that the greatest energy in the atmosphere would be generated late in the afternoon, when the sun had plenty of time to heat both the air and the Earth's surface. Strong winds flowing over the Rocky Mountains from the west would again mix with a descending pulse of cooler, drier air from the north. Those factors, along with the low-level moisture from evapotranspiration, provided enough turbulence in the atmosphere to generate strong thunderstorms capable of producing both large hail and tornadoes.

Our forecast maps showed two areas of Alberta with the greatest chance of being hit. One was between Calgary and Airdrie, and the second, with more potential for severe storms, was from Lethbridge southward toward the Montana border.

"Damn COVID," I said to Kyle and Leanne. "I wish we had two teams to cover these storms today."

Our company, like many others, was adhering to pandemic protocols to stem the spread of the coronavirus. One of the measures was no non-essential travel, so the luxury of having a couple of teams on one story was out of the question.

"Everything points to Calgary having an afternoon like Thursday and Friday," Kyle said. "The storms get big and are producing nasty weather, but they have a recent history of moving along fairly quickly, after a quick dump of hail and some funnels that have been weak so far."

"This area," he continued, "from the U.S. border near Del Bonita, then north to Taber and along Highway 36, looks like it has all the ingredients for some big, long-track tornadoes. The wind off the Rockies near Glacier Park in Montana creates much better vorticity as the storms form and then roll into Alberta. Calgary and Airdrie are definitely going to see hailstorms at around 6 p.m., but, as far as big tornadoes go, that's most likely to happen in that corridor from Warner to Taber to Vauxhall to Brooks, and could run on the ground for 60 kilometres or more, and they'll be impressive in size."

Kyle would drive south from Calgary toward the Montana border to intercept the storms that were forecast to form in that area during the late afternoon. We would rely on video from viewers and traffic cameras to document the storms in the Calgary area.

At our Oakville studios we updated the forecast with the severe storm watches that had just been issued. The weather was high-level overcast; it was not as warm as it had been on Friday.

In Calgary, Kyle packed up his truck with the equipment he would need to cover the storms: GoPros were mounted on the dashboard and roof, his laptops attached into specially built racks by the passenger seat, his camera and its batteries connected to chargers. At 2 p.m., he drove south on Highway 2 toward Lethbridge.

Two hours later, Kyle stopped in Lethbridge to check the weather models and current radar images. Some storm cells were showing up on his computer screen as small yellow patches with red centres, a typical signature of a thunderstorm on radar returns. They were southwest of Cardston, Alberta, and the cluster was forming up along the path he had discussed with us earlier in the day, just east of the Rocky Mountains. Cardston was an hour-long drive from his location in Lethbridge; the storms hadn't reached a mature stage yet.

Radar also indicates the direction and speed of thunderstorms. This grouping was travelling northeast at more than 25 kilometres per hour. His best option to intercept them would be to drive southeast on Highway 4 toward Stirling, Alberta. He could be there in thirty minutes, well ahead of the storm.

The flat cropland in this part of Alberta is crisscrossed with roads laid out in a grid, making it easy to follow storms across the landscape. Kyle pulled off Highway 4 and onto Highway 61, parking near the Richardson Pioneer feed elevators to the east of the town. He set up his camera on a tripod, the dusty Ford SUV with The Weather Network insignia emblazed on its side and a line of rail cars as the background for his report.

At 6 p.m., his thermometer read 24 degrees Celsius. The wind was coming from the south at 25 kilometres per hour;

2,000 metres above him, the wind was flowing much stronger from the west.

"There is really good vorticity available here in extreme southern Alberta, ideal to generate rotation and tornadoes," Kyle reported.

The internal workings of thunderstorms and supercell thunderstorms have been studied for years. Storm chasers take wind, temperature, and humidity readings as close as possible to the storms. The information they gather is critical to forecasting the behaviour of these often deadly and catastrophic systems.

The warm air rising in southern Alberta was being aided in its journey aloft by the southerly surface winds. As the warm air rises, upper-level winds streaming eastward from the Rockies begin creating rotation in the rising columns of air. Thunderstorms form, and their internal vertical wind flow draws warm surface air into its structure. In ideal conditions, like on this June day, a thunderstorm can take on mammoth proportions.

Now the storms roared to life.

Kyle watched the distinctive arch of a shelf cloud stretch across the horizon. An angry-looking yet beautifully structured inky—nearly black—mushroom-shaped cloud would pour out tens of thousands of litres of rainwater and an equally large amount of fist-sized hail. These hailstorms are often called "crop killers," as they are known for rolling over acreage at 80 kilometres per hour, crushing new growth.

After following Highway 61 east for 20 kilometres, Kyle drove north on Highway 36, over Chin Lakes, and 30 kilometres farther across flat farmland.

Dust kicked up behind the truck as he eased onto the shoulder just south of Taber. Nearly 9,000 people live in Taber, which is 50 kilometres east of Lethbridge. The city is noted for its sugar beet and corn production, and the fields that surround it supply local mills that turn the crops into consumable goods like corn chips and beet sugar. On average this area of Alberta gets 90 millimetres of rainfall during June, making it the wettest month of the year. Compare that to 50 millimetres in May and less than 40 millimetres in either July or August.

He could see debris clouds blowing upward from the base of a tornado as it moved over the fields. To the west the massive shelf cloud drew ominously closer to the town, travelling roughly to the northeast at nearly 60 kilometres per hour. It was 8 p.m. The storm had darkened the skies to twilight. Night would soon fall, and chasing the monster would become even more dangerous.

Rain and hail streamed from the giant shelf cloud as it passed overhead. Water pooled and flooded intersections as the tornado tracked around Taber, sparing the town serious wind damage. Kyle readjusted his cameras and continued to trail the storm northward along Highway 36.

As day became night, the only way to see the storm was to look skyward, where lightning continually illuminated the sky like a thousand strobe lights.

"I've never seen such a display before," Kyle reported. He had gotten ahead of the storm near Brooks. The tornado was now invisible in the night sky; only the massive, flashing storm was visible behind him. "The lightning is nonstop." The constant low rumbling of thunder was competing with his voice.

After tracking from Montana to Brooks, a distance of over

200 kilometres, the storm had finally begun to dissipate by 11 p.m. For storm hunters, nightfall usually signals the end of their chase; the sun's energy that helped fuel the storm and influence the wind is gone, and the atmosphere equalizes in the calm of night. This chase ended as the lightning diminished to occasional flickers while the storm drifted north across Red Deer River.

Kyle turned the truck and drove west, 190 kilometres back to Calgary. A mess was waiting.

Softly, a piano began playing the familiar notes that build to Paul Simon's words to "My Little Town." It was my phone alarm going off to this month's wake-up song. I let it play. When it began again, I picked up the phone and checked the time. It was 1:48 a.m. on Sunday, June 14, 2020.

I made two cups of coffee and placed them on the desk beside a dozen sheets of plain white paper and a red pen. I opened my emails and read through a document written by the nightshift producer at The Weather Network, sent an hour earlier.

There was massive storm damage in northeastern Calgary. I scrolled through my phone, clicking on videos of the storm. What played looked like scenes from a Hollywood disaster movie. Cars and trucks were submerged up to their windows, a sea of slush floating around them. Windows had been smashed, and it looked like a hammer had been used to hit the roof and trunk of a car—hail damage. The next video showed a typical suburban street. Vinyl siding hung in long strips from the nearly new homes; windows were shattered, and vehicles parked in the driveways looked as if they had been repeatedly fired at with a shotgun. Trees were bare of leaves; the ground was covered with hailstones.

As I drank coffee and took notes, I watched another video showing the enormous storm creeping toward the person taking the video, accompanied by a chorus of panicked and excited voices. Whoever shot the video started running back inside their home. I could see the sidewalk, then their feet as they jogged up three concrete steps, through a doorway, and onto tiles. In an instant the camera pointed back outside and toward the sky.

A tornado was descending from a massive black cloud. Children were running home, while a group of people stood in the middle of the street, shooting video of the funnel cloud that was writhing beneath the storm.

I scrolled to the next video. It was shot by a person walking down a cul de sac. The two-storey homes were stripped of vinyl siding, and the blue Styrofoam insulation was ripped out of the walls. "Jeez," I said to myself as I began reading reports from the *Calgary Herald*. The damage was catastrophic and didn't look typical of hail damage.

It was 2 a.m. in Oakville, midnight in Alberta. I called Kyle. He was still on the road, returning from his night of storm chasing.

"Have you seen the damage in Calgary?" I asked him.

"Yes," he said. "I'll go there as soon as it's light out." Daylight was five or six hours away.

I arrived at the office at 2:30 a.m. and began playing back the Calgary radars from Saturday afternoon. The forecaster who had been working that evening replayed the wind maps and it became clear to us what had happened.

Severe thunderstorm warnings had been issued for much of southern Alberta early in the afternoon, as they had been over the previous few days. Big cumulus clouds began to develop west of Calgary in the mid-afternoon. A brisk surface wind was flowing toward the higher terrain west of the city, providing orographic lift, which is the influence of elevated topography on wind patterns. The higher ground was forcing the moist warm air aloft. The clouds that continued to gather and grow were formed from water droplets condensing in the cooler air.

By 5 p.m. the once puffy clouds had become a massive, dark, ominous shelf cloud reaching several kilometres upward into the opaque sky. The base of this expansive shelf cloud seemed to hover a few hundred metres above the surface. The supercell storm was drifting eastward, pushed along by the strong westerly winds streaming over the Rocky Mountains. The 5:30 p.m. radar indicated that it would move over northern Calgary, likely giving heavy downpours along the Stoney Trail and Deerfoot Trail highways.

Slowly, the storm had enveloped northwestern Calgary. Rain fell in torrents; hail as large as golf balls peppered the ground; the wind grew stronger. Shortly after 6:30 p.m., Calgary International Airport reported its first rumbles of thunder; the hail followed soon after.

South of the airport, motorists driving at 110 kilometres per hour on Deerfoot Trail, the main highway through central Calgary, saw the first rain hit their windshields. The big drops fell sporadically at first. Then the sky opened above them, releasing a downpour. Even on the highest setting, the wipers couldn't maintain clear visibility.

Traffic slowed to 40 kilometres per hour. The soft and steady rhythm of rainwater was replaced by the dull pings of hailstones hitting the roofs and hoods of cars and trucks. The hail intensified and the stones grew larger. The pinging noises became thuds, accompanied by the sound of fracturing glass. In a matter of seconds traffic on the highways rolled to a stop.

Ice and water quickly overwhelmed the drainage system, and underpasses began filling with water. At the 32 Avenue interchange, the Deerfoot Trail became impassable; vehicles had stalled in the deep water and the route was closed.

As the storm rolled over the city, its forward motion slowed. The giant, slowly rotating supercell was suspended over the northeastern suburbs of Saddle Ridge, Skyview Ranch, Redstone, Martindale, and Taradale. The northwestern flank extended northward to Airdrie.

At 7 p.m., the weather station at the airport reported an abrupt wind shift; the powerful southerly surface winds turned 180 degrees, indicative of a strong downdraft on the western side of the storm. Storm chasers refer to this as a "rear flank downdraft." Winds were gusting at over 60 kilometres per hour and carrying the hail with it.

The hail was being slung at a 45-degree angle. Windows shattered. Siding was peeled away from buildings. Millions of hailstones as big as tennis balls were travelling at over 80 kilometres per hour, pounding away on thousands of homes, vehicles, and businesses. The noise was deafening. It seemed as if the storm would never end.

From the intersection of Métis Trail and Country Hills Boulevard, the base of the storm looked like a kilometre-wide ebony shaft releasing hailstones that came crashing to the ground. Piles of big hailstones, little hailstones, and ice pellets the size of peas covered people's lawns. Storm drains were clogged with ice, and the foliage had been stripped from the trees. The roads flooded.

The power flickered and then went out for over ten thousand customers. Traffic signals failed. Water flooded intersections and thoroughfares and blew open manhole covers.

Lightning strobed within the supercell storm, giving it the appearance of a giant alien mothership as it crept over Balzac and Airdrie, hugging the QE2 Highway as it moved north. The storm lasted ninety minutes.

As Kyle filed his report, the morning sun cast long shadows in the ravaged neighbourhood of CornerBrook. Behind him were battered cars and trucks with shattered windows. Houses had been stripped of most of their vinyl siding to expose bare plywood frames.

"For some of the people who live here, this is the second time in four years that they've had a storm do this type of damage," Kyle reported. "Insurers estimate the damage will be well over a billion dollars, making this the most expensive hailstorm in Canadian history."

"Albertans know too well the stress, turmoil, and financial hardship that severe weather events can cause," said Celyeste Power, vice president of the Insurance Bureau of Canada's western division.

By the end of that week, it was confirmed that the storm would cost over $1.2 billion to repair damages. More automobiles were written off in the province as losses than new cars were purchased.

With the addition of this storm, six of the ten most expensive weather disasters in Canada had taken place in Alberta, and all of them since 2010.

CHAPTER 12

Fire and Flood

British Columbia, 2021

Early in the morning on Saturday, June 26, 2021, three important events were impacting millions of people in Canada and the United States: the drought, the deadly heat, and the massive wildfires. There was no rain in the forecast for the next several weeks. Temperatures would continue to soar, guaranteeing that the wildfires would only grow larger. Linking the stories together was easy; explaining what led to these conditions was more difficult, given the complexities of climate science and weather behaviour, as well as the simple time constraints imposed by the television format. Each of the six segments per hour that I would create was just four minutes long.

My producer, Leanne Ferrante, and I met just after 4 a.m. Over cups of Tim Hortons coffee, we discussed the plan for the show and began assembling the maps, videos, and animations for each of the six segments. I headed into Studio A at five-thirty.

The room was large and obscenely bright. The air was cool, despite the racks of television lights hanging from the high ceiling. Black foam soundproofing lined three walls. The other wall and floor were bright green.

I stood in front of the green wall and in the centre of the green floor. A single camera was 5 metres in front of me. The camera would position me inside the virtual set. Television monitors were suspended from the ceiling on both sides of me and above the camera, just outside the lens's field of sight. The monitors showed the production as it looks on screen, allowing me to see where I was in the virtual world. They also displayed each item that was slated to play next in the story.

We would begin recording the segments about thirty minutes before they were scheduled to go on air; that way, if on the rare occasion we made a mistake, we'd have time to re-record. As soon as a segment was completed, we would send the four-minute block to the Master Control computer, which scheduled it to play at its allotted time in an automated sequence of local weather and commercials.

"Are you all set to do block one?" Leanne asked through the intercom.

"Sure am," I replied. The red numerals on an LED clock above the camera began a countdown to zero. At zero the computer played our opening animation.

In the monitors I watched the show intro and the clock start a new countdown to the end of this segment: 3:59, 3:58.

"I'm glad you're here. It's Saturday, June twenty-sixth," I said.

The image of a fire whirl played on a giant video screen behind me.

"Large forest fires create their own weather," I explained. "Can you imagine seeing this where *you* live?"

A fire whirl is a vortex created by an intensely burning fire. These eddies form when hot air rises. Dust and debris gathered by the rising column of air make visible the otherwise invisible vortex. Turbulent winds within a big fire create vortexes of hot air and combustible debris, which ignites to become a fire whirl or fire devil. Many fire whirls were observed during the incendiary bombings of Germany and Japan during the Second World War. When the magnitude of the fire is great enough, these vortexes can grow to the size and strength of tornadoes.

The first verified firenadoes occurred in brushfires near Canberra, Australia, in 2004. The winds in one were estimated to be moving horizontally at 260 kilometres per hour and rising vertically at 150 kilometres per hour. That firenado was consuming 300 hectares of brush per second.

More firenadoes were documented in California in 2018 and 2020. The National Weather Service estimated that the firenadoes observed in the deadly 2018 Carr Fire, near Redding, California, were EF3 in strength with winds of over 230 kilometres per hour. The Enhanced Fujita Scale is used to measure both wind speed in a tornado and the potential and observed damage. This system helps to better characterize the overall intensity of each individual tornado. Because of that particular event, the Weather Service began issuing fire tornado warnings in the summer of 2020 as a part of its forecasts in California.

In late spring 2021, forest fires had been burning for weeks in California, Arizona, and Nevada. It seemed all the U.S. Southwest was on fire, and now the wildfires in British Columbia were expanding and moving perilously close to communities.

The red LED numbers ticked away on the clock in front of me: 2:53.

In the massive virtual video screen beside me, red, orange, and yellow flames twisted against a black background of smoke. Trees were instantly incinerated as the spinning blast furnace passed over them.

It looked like a passage to Hell had opened on Earth.

Next the screen displayed an animated map that illustrated the spread of the drought over the past twelve months.

"This is the driest year since 1924," I said. "Less than half the usual precipitation has fallen since October 2020. It is the most extreme drought in modern U.S. history, and it has grown to encompass large areas of Mexico and western Canada."

A computer animation showed a viewer question from Twitter: "If the wildfires burn so much, why do the same areas burn again?"

"Forest fires are part of the natural cycle of forest regeneration," I replied.

Video of a wildfire burning in Oregon played on screen. "The heat of a fire helps to open seeds that will develop into the next season's new growth and the forest will begin to regenerate again, as it has done for millennia."

Two minutes, thirty seconds to go on the clock.

"Today our problem arises with changes to the climate cycle; drought has become more frequent and extreme in both magnitude and duration," I continued. "Global oceanic and atmospheric cycles over the Pacific Ocean have changed, resulting in deeper low-pressure systems and stronger high-pressure ridges that are impacting weather on the west coast of North America. The El Niño and La Niña cycles of warming and cooling in the Pacific Ocean have also fluctuated and become less predictable over recent years. That cycle has helped to drive the behaviour of these global weather patterns."

A map of western North America was on screen behind me; a big letter H, symbolizing "high pressure," slowly rotated over the west.

"Our hot, dry season in western North America is governed by a large cell of high-pressure, stable air that is slowly settling to the surface. The skies are clear, and heat builds in this area," I said. "That's why it is often referred to as a 'heat dome.' There is little to no circulation within the dome, just growing heat. This weather pattern is held in place by an atmospheric blockage, keeping weather systems nearly stationary for days and weeks on end."

I glanced at the clock. There were ninety seconds left. "New vegetation grows quickly with winter snowmelt and sporadic rain in the warmth of spring. However, as the drought reasserts itself in May, the plants become dry, grass goes dormant, and the land is ripe for fire."

Video of water bombers dropping red fire retardant onto blazing hillsides played on the screen behind me as I described how desperate the situation was becoming in California.

"Lightning, but more often humans, start the fires, which in this arid environment burn and spread rapidly. Already, the number of active fires in California has outpaced last year by 40 percent. Lake Mead, in Nevada, supplies water for Los Angeles and the farms of the Imperial Valley," I explained. "Its water level has decreased every year since 2000. Two of the state's largest reservoirs, Lake Shasta and Lake Oroville, are at their lowest levels in forty years."

I looked at the countdown clock. I had ten seconds left.

"There is no rain in the forecast. All the fires will grow, including those here in Canada, and the heat is coming. That part of our story is next," I said as the clock rolled to zero.

"I'll load up block two, and Kyle is on the phone," Leanne said through the intercom.

In the next segment, we were going to discuss a triple threat facing the west in the coming days. Smoke from wildfires was drifting over the mountains and impacting air quality in Alberta. For several days the smoke had blotted out the sun and had led to special health advisories.

It was 5:45 a.m. Leanne began the sequence to record our second segment. The LED clock counted down from five to zero and then began its new countdown from four minutes.

The Weather Network's "Active Weather" animation flashed on screen as video of forest fires burning near Kamloops, British Columbia, played.

"We are in for some really trying times in BC over the next few weeks," I stated.

The heat dome was expanding, circulating hot air across Canada. By Monday the temperature would climb by at least 5 to 10 degrees, from the low and mid-30s to well over 40 degrees Celsius. At the same time, smoke from wildfires that were burning in both Washington and British Columbia had begun to fill the valleys. The air quality would continue to deteriorate in the coming days and weeks.

Kyle was ready to go, and we brought him on screen.

"You spent years fighting forest fires; you breathed the smoke. How bad is the air quality?" I asked.

"It's bad," he said. He went on to explain how usually the prevailing westerly winds would help dissipate the smoke in Alberta, but with the lack of circulation under the heat dome, there was little movement of air.

"It's going to get worse in BC," he said. "The valleys are oriented north to south, and any chance of an upper air flow will blow west to east, so the smoke will be trapped in the valleys."

Video of the smoky red sunset in Calgary the previous evening played in the background as we spoke.

I stood on the virtual hardwood floor on the set, looking at the green wall and speaking to an imaginary video screen. Kyle was framed with the Calgary skyline behind him. He was standing on the balcony of his twelfth-floor condo, the camera on a tripod in his living room. The computer merged layers of video input; on my television monitor Kyle was inside the giant screen, I stood in front of the screen, and we talked about the next threat—the heat.

"It's very likely that the next five days will be the hottest ever in British Columbia," I continued. "This heat and the fires make it an exceptionally dangerous time for you. Where do you think the maximum heat will be?"

"Lytton or Lillooet, probably," he replied. "The hottest temperature we've ever recorded in Canada was back in 1937, during the Dust Bowl. It was 45 degrees Celsius in Saskatchewan. That record will probably be broken in the next week."

While we discussed the heat, a series of seven-day forecast animations played for various cities and towns in British Columbia. In Kelowna the high temperature was forecast to be 42 degrees Celsius; by Monday, it would be 44; and after that, even hotter. On the coast, the weather station at Vancouver International Airport would see soaring temperatures, up to 33 degrees Celsius by the end of the weekend. The forecast for Lytton was up to 48 by Tuesday. Weather models offered no relief; the "heat dome" would not ease this week, and it would be a part of the weather pattern for most of the summer.

As the clock counted down to zero, I thanked Kyle for his

insight and wished him well on his drive to the interior of British Columbia. He would report back to us throughout the week.

Heat warnings had been in place for several days. The Red Cross was pleading with British Columbians and Albertans to prepare and protect themselves. The 2018 heatwave in Ontario and Quebec left nearly a hundred people dead; the Red Cross warned that this heat wave would be longer and hotter.

"Block three next, with our health expert Rachel Schoutsen," Leanne said through the intercom in the control room.

For British Columbians, the temperature was suddenly 10 to 15 degrees warmer than normal, and in the temperate climate of the lower mainland, where the average daytime high in summer hovers around 22 degrees Celsius, 70 percent of homes didn't have air conditioning.

Rachel and I discussed the fact that all of the province's 5.2 million residents would be facing at least a week of extraordinarily hot weather and how important it would be to seek shelter from the sun.

A heatwave can be an insidious enemy. Heat drains moisture from our bodies, and slowly our internal temperature rises. That overheating can lead to serious and often life-threatening illness. Rachel and I discussed common-sense methods to keep oneself healthy.

"Avoid the midday sun, and try not to exert yourself," Rachel said. "Skip any extraneous activities, drink lots of water, and avoid alcohol and stimulants."

Video of people sunning themselves on Kitsilano Beach played on screen.

"It's important to keep air circulating if you don't have AC," I added.

"It is," she replied. "Use fans and keep your windows open.

Take cold showers or sit in a cold bath; that will help keep your internal body temperature down. It's also really important to check on each other, to make sure we all stay safe this week."

The clock continued its countdown toward zero and the shot changed to a two-box, showing Rachel sitting in her kitchen and me standing in the virtual studio.

"It's uncomfortable to talk about how deadly this has the potential to be," I stated. "But the last significant heatwave for British Columbia was in 2009, and the coroner's office estimates that 110 people died because of the heat that year."

Rachel nodded and added, "A study of that 2009 heatwave in the *BC Medical Journal* noted that there was a higher rate of mortality among the elderly and those with chronic illness, but also a noticeable surge in sudden deaths among younger people because they had overexerted themselves. It's so important to take this seriously."

It was just past 6 a.m. We'd produced the first half of our show, and the very first segment was just going on air as we prepared to record block number four. As was usually the case, the production was coming together without any problems.

Tyler Hamilton, one of our meteorologists who specializes in western Canada, would review the ongoing wildfires in BC and what the next week might hold with the extreme heat.

He joined me from Studio B, which was identical to Studio A. COVID protocols prohibited us from sharing the same workspace. The computer assigned Studio B a different background set and camera angle from what we were using in my studio. Then Leanne began the countdown clocks for both of us.

A "BC Wildfires" animation played and then cut to video

of blazes burning near Peachland, just southwest of Kelowna. Flames leapt skyward from exploding ponderosa pines as a Canadair CL-215 water bomber dropped its load on the fire. Five thousand litres of water fell from the sky in a white plume on the burning embers.

"This fire season started early, and it's going from bad to worse, isn't it," I said as the video of the firefight continued to play.

"It is," Tyler replied. "For some context, in 2017 over 12,000 square kilometres of forest burned in BC; in 2018 the number grew to over 13,500 square kilometres. This summer is going to be drier and hotter, so you can see where this has the potential to go."

An animated map that illustrated the expanding drought over the coming sixty days showed surface moisture rapidly depleting by late July. Another animation followed, showing the wildfire risk quickly elevating from high to extreme over the next four days.

"The forests of British Columbia are a tinderbox under the heat dome this week," Tyler explained. "The fires that are burning now will continue to grow, and any new fires are going to be difficult to gain control of. For years, climate scientists have said the weather and fire conditions that we've been watching in California for the past twenty years will become part of our reality in BC. This week is proof that the science was right, and it is happening here now."

Tyler is from Campbell River, British Columbia. The city of 35,000 sits at 50 degrees north latitude, at the south end of Discovery Passage on the east coast of Vancouver Island. He knows the importance of the forests, both economically and environmentally.

"The fires in 2018 burned an area nearly three times the size of Prince Edward Island," Tyler continued. "Or for more perspective, if you're from BC, an area the size of Vancouver Island, the forty-third-largest island on Earth, has been burned over the past four years."

We continued to discuss the immediate implications of our changing climate. Individual weather events were becoming more extreme in their nature and were happening with greater frequency. In British Columbia, our ability to adapt to the latest weather event—extreme heat—would be tested in the coming days.

The clock was winding toward zero again. I asked Tyler, "What might be a worst-case scenario this week?"

He mused for a second and then replied, "A fire near a community in one of the valleys or canyons in the interior. No doubt, that could be catastrophic on many levels."

We continued our production as Environment and Climate Change Canada reissued heat warnings across western Canada.

"This hasn't happened before," I said as the new advisory map filled the screen next to me. "All of BC, Alberta, Yukon, a huge swath of the Northwest Territories, Saskatchewan, and Manitoba are under warnings for extreme heat."

The computer displayed the seven-day forecast map for the week ahead. The colour on the map shifted back and forth between shades of orange and red.

"The red indicates temperatures over 40 degrees Celsius during the day, but look at the evening and overnight readings—they rarely drop below 25 or 30 degrees. There is literally no break from this relentless heat," I said. "If we achieve these extremely high temperatures for the duration that is forecast, it will make this week unprecedented in Canadian weather."

That weekend the heat dome continued to strengthen over western Canada. The centre of high pressure was over British Columbia, and the downward pressure of the atmosphere served to further heat the air near the surface. It was as if the heat dome was acting as a pressure cooker for the air within it.

On Sunday, temperatures were already surpassing the forecast. Vancouver Airport reported over 31 degrees; downtown recorded 33. The interior was hotter. Kamloops registered 42.8 degrees, Kelowna was nearly 42, and Lytton set an amazing new record.

I have travelled to Lytton several times. It's a small village located in a deep, narrow valley where the Thompson and Fraser rivers merge. The Nlaka'pamux People have lived in the area for ten thousand years and call the place Camchin (ƛ'q'əmcín), which means "river meeting." Europeans arrived in 1858, during the Fraser Canyon gold rush. They called the area The Forks, though the name would soon be changed to Lytton in honour of British colonial secretary Edward Bulwer-Lytton. If his name somehow seems familiar to you, perhaps it's because he was also the author who began a novel with the infamous words "It was a dark and stormy night." The village was established on the main transportation route from the Pacific coast to the interior of British Columbia and the rest of Canada. The railway came in the 1880s, then the Cariboo Highway in the 1920s and the Trans-Canada Highway in the 1950s.

Until 2007, the forestry industry was the village's main employer. Today there a few service stations and restaurants for passersby, but Lytton is now best known for consistently

racking up new high temperature records for both the province and the country. It's also the self-proclaimed "River Raft-
ing Capital of Canada."

Outside the library in the village centre, the streets were
lined with black locust and Manitoba maples. It was dry
and dusty, and the sun glared from a clear blue sky on most
summer days. In the shade, the thermometer regularly reads
35 degrees Celsius from late May to mid-September; less than
60 millimetres, or 2 inches, of rain falls in an average summer.

Only about 250 people live in the actual village of Lytton,
but beyond Main Street the surrounding community adds
another 1,500. It's a place where people know each other; you
can leave your car unlocked with the keys in the ignition when
you stop at the village office to pay your bills.

Kyle arrived in Lytton on the morning of Sunday, June 27.
The temperature was forecast to reach 46 degrees Celsius,
making it the hottest day ever in Canada. Already records
had been set across western Canada for the warmest day in
a particular town or city, but this was big—this would be the
hottest temperature ever recorded in our country.

A few weather watchers went to Lytton too. They sat in
the square, glued to their thermometers as the mercury soared
past 41 degrees at noon. By two in the afternoon, the threshold
had been surpassed and the mercury continued to climb, up to
46.2 degrees Celsius by 4 p.m. The old record of 45 degrees,
set eighty-four years earlier in Saskatchewan, had been broken.
Lytton would now be known as the hottest place in Canada.

Kyle reported on the celebratory mood in the little village
as the sun dipped lower in the western sky and the thermometer gradually ebbed back toward the high thirties. It was blazing hot, and the next day would be even hotter.

People shared photos and memes about the extreme heat on social media. Kyle cooked bacon and eggs on the roof of his truck, then cooked up another serving on the asphalt of a village street. People timed how long they could walk barefoot on the pavement.

On Monday afternoon the new records for heat that had been set on Sunday afternoon were surpassed. Lytton was 47.6 degrees Celsius. In Kamloops it was 43.3, Kelowna recorded 41.8, and in Vancouver, where the average summer high temperatures are between 20 and 22 Celsius, the mercury soared to 33. The heat was now four days old. Tuesday was forecast to be hotter still.

The oppressive dry heat was drawing moisture from all living things. Forest fires were spreading much faster and containment efforts were becoming futile. On Tuesday afternoon a large wildfire had started near Sparks Lake, just northwest of Kamloops. The number of fires had now nearly doubled the seasonal average to date.

On Tuesday, June 29, the heat peaked. The high-pressure cell that had been holding the heat dome over the west was creeping eastward. Over the next forty-eight hours the wind would shift slightly to become more southerly, aligning with the interior valleys; it would also increase in speed. Then the relentless, record-breaking heat would begin to wane.

At 3 p.m. the mercury stopped climbing in Kamloops. It was 47 degrees, their hottest day ever; Kelowna reported 44.8.

In Lytton it was 49.4 degrees Celsius.

Not only was that the highest temperature ever recorded in Canada, but it was also the highest temperature recorded anywhere on the planet north of latitude 45. The temperature was the highest recorded in North America outside of the

U.S. Southwest desert, and hotter than any reading ever taken in Europe or South America.

That day new records were set in British Columbia, Alberta, Saskatchewan, Manitoba, Ontario, Northwest Territories, Yukon, Washington, Idaho, Montana, Oregon, and California.

Climate scientists explained that this once in a once-in-a-thousand-year heatwave was made 150 times more likely by global warming. There was no doubt that this extreme weather event was a product of rapid climate change.

On Wednesday, June 30, down-sloping winds blew into Lytton from the south. The air was hot, but not as dry as it had been before. By afternoon the temperature was 38 degrees, though the humidity made it feel like 44. The wind was strong and steady at over 40 kilometres per hour.

Sometime just before 6 p.m., a fire started. The ignition might have been caused by a random lightning strike or a hastily discarded cigarette; perhaps it was a spark from an ATV or from the passing of a 157-car freight train. There had rarely been an environment that was so conducive to a fire.

Flames raced northward, pushed by the strengthening wind, up the valley toward the village. The roadside's dried bunchgrass and the fir trees easily ignited and burned. The flames leapt toward new fuel, sparks were lifted by the wind, and the main blaze set everything alight.

Volunteer firefighters worked desperately to contain the fire. That effort bought the community a few extra minutes to run for their lives as structures just south of the village were quickly engulfed in flames.

Propane tanks exploded as the fire raced through a gas station. Flames devoured everything in the widening path of the firestorm. Word quickly spread that the fire was uncontained and moving toward the village at 20 kilometres per hour.

Mayor Jan Polderman ordered an immediate evacuation. It was 6 p.m. The inferno was less than 15 minutes old and raging into the community. There was no time for residents to gather anything. People simply got into their vehicles and drove north. Only two routes led away from the inferno: Highway 12 to Lillooet or the Trans-Canada toward Spences Bridge. There was little time for an organized exit.

In less than an hour, 90 percent of Lytton had burned to the ground. The fire destroyed nearly every structure in the village, including all the buildings on Main Street as well as the museum, library, and village archives. The price of the firestorm was far higher than the $78 million loss assigned by the Insurance Bureau of Canada.

As a wall of flames raced down on their home, Jeff Chapman helped his mother, Janette, and his father, Mike, take shelter in a trench outside their house. Only Jeff knows the terror that he and his elderly parents faced that day. The inferno did not pause or consider human life as it sped through Lytton; it only consumed.

Jeff hastily threw a sheet of plywood over the trench to protect his parents from the flames as the fire engulfed their house and raced toward him. The heat was unbearable and he turned, running away from the fire and air that had become too hot to breathe.

He ran toward the river; it was the last time he saw his parents alive.

Fire experts drew parallels between the 2018 Camp Fire in California and the Lytton Fire. Both incidents saw communities obliterated by a fast-moving firestorm. The California fire was sparked by a fire on an electrical transmission line and grew to engulf 621 square kilometres. A firestorm erupted and overtook the towns of Concow and Paradise. It took less than an hour to burn down a town. A steady 90-kilometre-per-hour wind acted as nature's bellows, feeding oxygen directly into the heart of the blaze. There was only a moment's notice for the 26,000 Paradise residents to evacuate. Eighty-five people died as the fiery maelstrom turned the community to ashes. In an hour, 18,000 structures were destroyed, and Paradise was lost. The fire was the deadliest and most destructive in California's history.

During the summer of 2015, I spent a week reporting on wildfires that were burning near Pemberton, British Columbia. The steep canyons that rise nearly vertically are lined with fir, pine, cedar, and hemlock trees. It is one of the largest unlogged tracts of forest in southern BC. In hot, dry conditions these valleys become extremely combustible. Topography aids the ease with which a fire can grow. Flames reach upward into the trees growing on the canyon's walls. Heat from the combustion creates strong winds and updrafts that push a fire across the forest canopy.

At Pemberton Regional Airport, we watched helicopters and water bombers fly countless sorties every day, trying to contain a large fire burning west of Pemberton. The aircrews were from across Canada, the United States, and even New Zealand and Australia. Aerial teams worked long hours to slow the fire's progression so that ground crews could enter the forest on foot.

On the ground, firefighters worked in 40-degree heat that was made hotter by their heavy fireproof clothing. It was a job done one square metre at a time. Smoke filled the air, and hot embers showered them while they used shovels to beat flames into submission. It is extremely dangerous work.

There is a coordinated method of attacking a blaze. Every morning at sunrise the teams would assemble at the Pemberton Fire Base adjacent to the airport. A detailed weather consultation was conducted. Wind behaviour was the item of most interest, because the wind would dictate which way the fire would move across the forest. The day's fire plan was reviewed, adjusted, and then implemented. Crews knew exactly where to find weak spots in the fire structure; they would attack those areas in an effort to cripple the fiery beast.

The air and ground crews worked seven days a week, sunup to sundown, until their rotation was over. After a few days of rest, they would be back on the line. As more fires developed, many of the aircraft and their crews would move elsewhere in the province to fight on those new fronts. Fire season usually extends from the first fires in April until the last embers are extinguished in October.

On June 30, 2021, some of those same crews would be working the two new blazes that were exploding near Lytton and near Kamloops at Sparks Lake.

As Lytton burned, Kyle was 150 kilometres away, driving along Seven Lakes Road on the eastern side of the juggernaut that was the Sparks Lake fire. His experience fighting wildfires led him to that route; it was the safest way to get close to the blaze and the safest way to get away from it, too.

A massive pyrocumulus cloud towered over Sparks Lake. The pillar of grey and white smoke formed a mushroom shape 18 kilometres into the sky. Lightning flashed and streaked within the cloud; dull rumbles of thunder accentuated the low drone of the all-consuming fire.

Birds flew past the truck as Kyle opened the door. They made noises he'd never heard before. The thousands of animals that made this forest home were in a panicked flight away from the monster that was devouring their world.

That night the roaring wildfires created their own weather system. The sky was alive with pulses of lightning. An unprecedented 710,000 lightning strokes were recorded in British Columbia that night. Many of those flashes would connect with the dry forests and ignite new fires.

The Sparks Lake fire would grow to become the largest in the province. By early September the fire was nearly extinguished. It had consumed 900 square kilometres of forest and forced the evacuation of over a thousand properties. An investigation linked its ignition to a suspicious SUV fire at an illegal marijuana-growing operation near the lake.

The Lytton Fire would continue to burn for weeks, eventually charring 837 square kilometres of land.

The government of British Columbia issued a province-wide state of emergency in July because of the extreme wildfire risk. The heat dome had done its work.

Between Friday, June 25, and Thursday, July 1, Canada Day, at least 595 people died because of the heat; 231 of those people lost their lives in one day—June 29, Canada's hottest day ever.

On July 13, lightning started a blaze near Whiterock Lake. The fire was 10 hectares in size when it was spotted in the late afternoon. It took thirty minutes for firefighters to attend. In that time the fire had tripled in size and flames were raging through the canopy of cedars.

The dry forest was like gasoline.

Within three weeks that wildfire had destroyed over 300

square kilometres. Thick smoke filled the valleys of the Okanagan, Thompson, and Nicola regions. The daytime sky looked like dusk as the choking smoke stubbornly blanketed the communities of Kamloops, Vernon, Armstrong, and Salmon Arm. Grey ash and embers fell continuously from the sky for weeks.

By the time August arrived, evacuation alerts had been issued for dozens of villages and towns in BC, including for the 4,800 residents of Armstrong and the 40,000 people who lived in Vernon. The hamlet of Monte Lake was burned to the ground, and homes went up in flames in Ewings Landing and Killiney Beach.

Throughout the summer, smoke from these fires and countless others in both British Columbia and Washington drifted westward down the Fraser Valley. Guided by weak winds, a pale of smoke hung over Vancouver. The usual blue summer sky had been painted in shades of grey and beige; visibility was often reduced to only a few kilometres.

On Friday, August 13, Vancouver had the worst air quality of any major world city. A reading of 200 is considered unhealthy; Vancouver registered 256, meaning very unhealthy. Kamloops and Kelowna were both over 300, which meant breathing the air was hazardous to your health.

If only the weather pattern would change; if only it would rain . . .

Rain would help douse the forests and the fires would become more manageable.

Rain would clear the atmosphere of particulate—the ash, dust, and dirt—and help cleanse the sky.

Rain would extract some heat from the atmosphere—the air always cools when it rains.

Finally, it did rain, but not when it was most needed.

Instead it was part of the usual autumnal weather pattern that develops along the Pacific coast. And in a year of extremes, the worst was still to come in British Columbia.

If you've looked at a satellite image of our planet, you will have seen an atmospheric river—a long strand of wispy clouds moving toward the poles from the equator. These moisture-laden clouds carry nearly 90 percent of all the water vapour that is transported in the atmosphere at mid-latitudes.

The rivers of moist air in the atmosphere have been assigned names to help identify where they originated. "Pineapple Express" is the term often used for an atmospheric river that begins its journey near the Hawaiian islands, arriving on the mountainous west coast of North America during the autumn and winter months. The name of the air mass reminds us of the tropical humidity in warmer climates.

Every year, twenty-five to thirty atmospheric rivers spill their contents along the coast of British Columbia. The inbound moisture, contained in a layer of cloud hundreds of kilometres wide, meets the mountains and is hoisted farther aloft, where it cools and condenses. Precipitation in the form of rain or snow or both, depending on the temperature and elevation, will fall, usually heavily and for several days.

These rivers are directed toward the west coast—from California, where they provide half the state's annual rain, northward to the Alaska Panhandle—by deep areas of low pressure that climactically form in the Gulf of Alaska at that time of year. The powerful upper atmospheric wind known as the jet stream forms a barrier between the strong low pressure near Alaska and high pressure further to the south. The atmospheric river is spirited along toward the west coast by the jet stream.

Often the powerful winds of the jet stream will mix downward toward the surface, making the arrival of an atmospheric river all the more dramatic with gale-force wind. It is not uncommon for wind speeds to reach more than 160 kilometres per hour on the west coast of Vancouver Island and farther north in Haida Gwaii.

After a hot, arid summer, the first in a series of atmospheric rivers began to soak British Columbia in mid-October. The autumn weather pattern had finally arrived. The much-needed rain extinguished the last of the summer wildfires, and in the mountains, above 1,000 metres elevation, heavy snow fell.

Over three days, from October 14 through 16, Vancouver reported 121 millimetres of rain; 153 millimetres fell in Abbotsford; and Chilliwack reported 178 millimetres of rain. The parched earth was thirsty for moisture. More rain fell intermittently in the following week, 25 millimetres one day, then 40 millimetres a few days later.

Most days in late October and early November were cool and cloudy. Without the evaporative power of the sun, the moisture from the persistent rain showers over the past five weeks had saturated the ground. In October, half of the month's thirty-one days had seen rainfall in Vancouver and the Fraser Valley. November recorded rain on twenty out of thirty days.

On November 5, computer weather models verified a strong signal that a potent atmospheric river would impact the west coast. The next series of storms could result in 250 millimetres of rain in one twenty-four-hour period. That's 10 inches of rain, an almost unheard-of figure for BC.

The atmospheric river would begin producing precipitation

on Thursday, November 11; slowly at first, then a steady rain that fell throughout the day on Saturday. Downtown Vancouver reported 17 millimetres of rain; across the harbour, North Vancouver reported 39 millimetres. On Sunday the relentless west coast downpour continued. Vancouver Airport was reporting a steady 60-kilometre wind from the west. Rainwater pooled on the ground and ran like rivers across intersections.

From its origins in the tropical Pacific, the 4,000-kilometre-long stream of moisture-laden air was 500 kilometres wide when it hit the coast, extending from northern Oregon to northern Vancouver Island. It was as if a firehose had been opened and directed at the coastal mountain range.

At 1 p.m. on Sunday, the provincial highway and transportation emergency media service, DriveBC, warned people to avoid all unnecessary travel in southern British Columbia and into Washington.

The summer wildfires had altered soil makeup in vast areas; the superheated air of the fires had turned the top layer of ground hydrophobic. Beneath the ash and cinders, this hydrophobic surface would repel water rather than absorb it. The waterlogged layer of post-fire debris would simply slide off the sides of the charred canyons. Mudslides.

The warm rain was also melting the metre-deep snowpack above the alpine level in the mountains. That meltwater, as well as the abundant rainwater, was cascading from higher elevations. Creeks and streams became raging rivers, and waterfalls tore away at the banks and surrounding vegetation as they raced toward the valley floors. Boulders as big as cars were loosened by the erosive precipitation and cascaded down the mountainsides, along with trees, mud, and stone. As the top layers of ground were stripped away, the endless rivers of water flowed more rapidly into the valleys.

On Sunday evening, families driving from the Okanagan to Vancouver on the Coquihalla Highway were stopped in their tracks by a highway littered with boulders.

Near Mission, a tsunami of mud rolled off a mountainside and across the highway. More than a dozen vehicles were pushed off the road by the mudslide. The occupants would spend the cold, wet night stuck in their cars as water and debris flowed around them.

The Tulameen and Similkameen rivers raged into Princeton, tearing away roads and infrastructure as the water rose and filled basements.

Rain continued to fall relentlessly on Sunday night. The floodplain that extends west from Chilliwack to Vancouver began to fill with water from the Fraser River and its tributaries.

The Lower Fraser Valley is accustomed to floods, particularly in years with deep winter snowpacks and heavy spring rains. The second-worst flood occurred in 1948, when the Fraser River rose 7.5 metres above its banks and left 200 square kilometres of land underwater. The discharge rate during that flood was 15,600 cubic metres per second; the average flow is 3,400 cubic metres per second. The 1894 spring flood was worse: it saw the Fraser rise 7.8 metres above its banks, and the discharge rate was estimated to be 17,000 metres per second. A huge inland sea formed, and the main CP rail line was washed away.

A century ago, Sumas Lake and its extensive wetlands lay between Abbotsford and Chilliwack. To mitigate future spring floods, the lake and adjacent marshlands were drained, and a series of canals and dikes were constructed to manage water flow in the valley. That "renovation" of the landscape helped to create the rich agricultural land that is our modern Lower Fraser Valley. Three hundred thousand people live there, servicing the agricultural sector that thrives on the rich floodplain and helps feed the 2.5 million inhabitants of metro Vancouver just 40 kilometres away.

Near Langley on November 14, 2021, the glare of red brake lights defined the eastbound traffic that crept along, side by side, on the four-lane Trans-Canada Highway. Heavy rain made visibility nearly zero as wipers raced to keep up with the sheets of rain that strafed across windshields; road spray and the darkness of night compounded the problems.

Outside of Abbotsford, the flashing red lights of RCMP cruisers brought traffic to a standstill. Out the window where fields should be, there was only the blackness of water.

It was 9 p.m. when the mayor of Abbotsford, Henry Braun, issued a local state of emergency. Floodwater was overtaking the city of 140,000. The Nooksack River, 10 kilometres to the south in Washington, had overflowed and flooded the flat fields between Everson and Abbotsford. Water spilled northward across farm fields to reach the Trans-Canada. Soon the highway would be underwater too.

Police officers stood in the teeming rain; water swirled around their ankles as they redirected traffic back toward the Vancouver metropolitan area and tried to explain the situation to unbelieving motorists.

Ninety kilometres to the east, in Hope, the highways were closed in all directions. Over a thousand travellers were stranded in the little town; for those who had perilously navigated around the fallen boulders on the drive from the

Okanagan, mudslides and washouts had finally closed Highway 3 from Princeton behind them. On Highway 5, the Coquihalla, the entire roadbed had washed away. The main highway to Merritt and the Okanagan was closed. Highway 1, the Trans-Canada to Lytton, was impassable due to landslides and erosion. The marooned motorists couldn't return eastward or to the north, and now overland flooding and the threat of mudslides had closed the highways that ran westbound toward the coast. Twelve hours after the travel warning had been issued by DriveBC, there were few places you could drive to.

More worrisome, the gravel beds along the major rail lines in the Fraser Valley were eroding. Near Yale, washout and the derailment of a freight train effectively closed the main line through the mountains.

Natural gas and petroleum pipelines were shut down as the floodwater continued to rise on Sunday night. A ruptured pipeline would be an ecological disaster.

The flooding rain and accompanying strong winds also disrupted the electrical grid. At one point over the weekend, nearly 300,000 customers were without power.

By the time the torrential downpours began to ease to a steady rain on Monday, the records for a one-day rainfall had been smashed. The weather station at North Vancouver estimated 288 millimetres; downtown Vancouver measured 110 millimetres; and Victoria checked in with 120 millimetres of rain. To the east in the Fraser Valley, weather stations at Pitt Meadows and Abbotsford each reported 160 millimetres, Chilliwack measured 195 millimetres, and Hope reported 242 millimetres of rain in one day. Many regions had received nearly a foot of rain. The floods would have happened anywhere considering the volume of water that poured out of the sky, but the topography of British Columbia's Lower Mainland and Fraser Valley combined with the extreme weather during the previous six months had laid the groundwork for the perfect natural disaster.

It rained all day on Monday and the water continued to rise. Road and rail lines were closed; Vancouver was cut off from the rest of the country. The Port of Vancouver, our nation's busiest, was forced to halt operations.

In the Fraser Valley, school was cancelled. Students, teachers, and parents—everyone was filling sandbags to reinforce the dikes. Pumps that drain the canals encircling Abbotsford and Chilliwack were running nonstop to stem the rising water; there was a growing concern that the pumps could break down. Volunteers and emergency services spent the day rescuing thousands of people from their homes. The Red Cross, with the help of dozens of service groups, began the hard work of feeding and finding accommodations for the displaced.

Two hundred kilometres to the north, rising floodwaters forced the evacuation of all seven thousand residents of Merritt. The usually dry, dusty town, known as the Country Music Capital of Canada, was now underwater.

Just outside Lillooet, a mudslide wiped Highway 99 from the side of a mountain. Several vehicles were swept away, and at least four occupants were killed.

The atmospheric river had now officially turned deadly.

On Tuesday the rain began to ease, but the water continued to rise. The Fraser River has its source at the Fraser Pass on the Alberta–British Columbia border near Mount Robson. The river flows 1,375 kilometres to its delta, just south of

Vancouver, where it discharges water at a rate of 3,400 cubic metres per second. It's the longest river in the province and eleventh longest in Canada; the river and its tributaries drain over 220,000 square kilometres of land in southern BC.

The sound of helicopters reverberated in the sky as they carried out repeated search-and-rescue missions over the flooded valley. All manner of watercraft had been pressed into use, ferrying residents who had been left stranded by the floodwater to higher ground. Fifteen thousand people were now temporarily homeless.

Barns and livestock sheds were submerged. Over 640,000 chickens, cows, pigs, and other farm animals had drowned so far; the number of dead livestock was equal to the number of animals lost to the summer heatwave. Farmers travelled in boats assessing the damage. Fields that had offered a rich harvest of berries and vegetables were submerged under metres of water. Toxins from nearby industrial lands, gas stations, and waste dumps were leeching into the water.

Sumas Lake had been reborn, larger than it had been a century ago. The lake extended southward from the Fraser River, beyond the international border to reach the western outskirts of Bellingham, Washington. The lake was being filled by runoff from the Nooksack, Chilliwack, and Fraser rivers. This is how the natural floodplain has worked for millennia. Our modern towns were now isolated islands that could only wait for the water to recede.

That week, fuel rationing began in southern British Columbia. With pipelines submerged and shut down, fuel had become scarce.

Commodities that usually flowed through Port of Vancouver sat on forty-two moored ships waiting to offload cargo.

Those ships would then reload with the raw materials that sat in the train cars stranded east of the washed-out rail lines. The storm had temporarily, but severely, interrupted the economic supply chain.

Above, the atmospheric river was waning. The deep area of low pressure near Alaska had begun to stabilize. This initiated the return to an equilibrium in the atmosphere, but the aerial sea was in a state of perpetual change. The atmospheric pendulum would swing again, allowing for favourable conditions to repeat the storm cycle. Precipitation and strong winds would return to the coast; the location of the jet stream would determine where the next storm would hit.

In the weeks that followed, the water would recede and the roads would reopen. Gas flowed through pipelines and the railbeds were repaired. Insurance adjusters estimated the damage at a half-billion dollars in insured losses, but much of the loss on the floodplain was uninsurable. Flooded homes and businesses and agricultural and commercial losses would continue to grow over the final weeks of 2021.

The disaster became our nation's most expensive; estimates suggest a cost of over $7.5 billion. That number does not include the summer fires or heatwave.

It was the world's fifth-costliest climate disaster in 2021, behind only the $15 billion floods in China; a Texas deep freeze that cost $23 billion; spring floods that led to $43 billion in damage to Germany; and Hurricane Ida, which caused an estimated $65 billion in damage to the United States.

Every year the price tag for damage directly related to our changing climate rises, and every year unmanaged or ignored threats need the convergence of only a few unfortunate circumstances to create a disaster.

In the Fraser Valley floodplain, the water is gone, absorbed into the rich earth. But as of 2022, a look just beneath the surface will reveal that the soil is saturated with moisture and awaiting the coming annual cycle of spring rain and snow-melt.

EPILOGUE

In February 2004, I sat on a stool in the rotunda of the Royal Ontario Museum in Toronto. The space was lit with four racks of television lights; two cameras were pointed at me, and a third was pointed at the audience.

Fifty grade four students were sitting cross-legged on the floor. Their eyes were wide and mouths were hanging open as they looked up at David Suzuki, who was sitting beside me. We were talking about climate change.

The subject matter and discussion were beyond their level of comprehension and attention span. For the past fifteen minutes, many of the children had been rolling on the floor and amusing themselves with plastic dinosaurs, their teachers asking them to hush when the chatter grew beyond whispers.

But a minute earlier, David had said something that caught their attention, and they were now hanging on his every word.

"The Arctic is warming at least twice as fast as anywhere else on the planet," he said. "When ice melts and the permafrost begins to thaw, huge amounts of methane will be released. That will lead to warmer temperatures, and the ground will melt even faster."

Then he added, "Scientists don't know what undiscovered germs and viruses from hundreds of thousands of years ago are frozen in that biomass."

Despite their age, those students understood the danger in a warming Arctic.

In that moment, it felt like we'd stolen some of their innocence.

Today, all of us understand the profound impact of a new germ or virus. The COVID-19 pandemic opened our eyes to the delicate balance that exists in the natural world and to the fragility of our own existence.

Over the years, David and I have had several conversations about climate change and the clear signs that the weather is becoming more severe because of the rapid warming of our planet. These climate changes, caused by human industry, are a threat to all of us on many levels.

David Suzuki can be a polarizing figure and has been chided for being overly rigid in his assertion that we must act now. But David is simply the messenger for climate scientists around the world; their decades-long studies are backed by facts and indisputable data.

A warmer atmosphere holds more moisture, so heavy rainfall events and flooding are more frequent. Warmer oceans provide more energy for hurricanes, typhoons, and tropical cyclones; they too have become more common and severe. As global weather patterns shift and change, heatwaves and drought also become more widespread. Cold snaps, though less frequent, become more profound in their deviation from what we consider to be average. The United Nations and governments around the world agree that the planet is at risk. Already there are ongoing mass extinctions of epic proportions.

In 2022, the Intergovernmental Panel on Climate Change presented its most detailed look at threats posed by global warming. Authored by 270 researchers from 67 countries, the report was approved by 195 governments around the world and states that the planet is "being clobbered by climate change." In it, scientists say that many of our efforts to adapt to changes are incremental and that we are not making the transformational changes necessary to stave off irreversible threats to entire ecosystems.

The largest living structure on Earth, the Great Barrier Reef, is dying. The reef covers 344,000 square kilometres in the Coral Sea and sits 100 kilometres off the coast of Queensland, Australia. More than half of its coral organisms have died since 1995. They are simply unable to adapt quickly enough to the warmer sea temperatures; scientists say that it is a massive natural selection event happening right before our eyes.

When the reef dies, an entire marine ecosystem will collapse. A habitat for sharks, sea turtles, crustaceans, and thousands of species of fish will change; new predators will arrive, further disrupting the delicate natural balance in this part of the world. Our great oceans, which cover 71 percent of the planet, provide food for over 3 billion people. The warming of the oceans is making them uninhabitable for thousands of aquatic species.

The Amazon Rainforest is the world's largest, covering 5.5 million square kilometres in South America. It is the most biodiverse place on our planet, with 16,000 species of trees and 40,000 types of plants, 2.5 million insect species, and over 4,000 species of reptiles, fish, amphibians, birds, and animals. Changing weather patterns that lead to drought and wildfires, as well as clear-cutting of the forest for industry, are putting this region in peril. Two consecutive years of drought, in 2005 and 2006, were the worst in a century; more droughts followed in 2010, 2015, and 2016. Research undertaken by both the Woodwell Climate Research Center and the National Institute of Amazonian Research confirms that the entire rainforest is being pushed toward a tipping point. Specifically, three or four consecutive years of drought coupled with human deforestation could turn the rainforest into a vast savannah or prairie.

If we lose the Amazon Rainforest, nearly half of the world's plants, animals, and microorganisms would die off. Nearly 90 percent of the medicines we use are derived from plants that grow in the rainforest, and it holds 20 percent of the planet's freshwater. As well, the rainforest stores carbon dioxide, a greenhouse gas; its destruction would impact air quality and further disrupt global weather patterns.

David and I have often spoken at length about the geopolitical side of climate change.

"A billion mouths need food in China," David said, "a billion more in India. Both of those nations grow their food on land that is highly susceptible to flooding by rising sea levels. Drought is providing wide-scale famine in Africa; the Sahara Desert is expanding because of changing climate patterns."

Nations go to war to feed themselves—a sobering thought.

The last time we spoke, in 2019, he said to me, "I don't know why we're even having this discussion. No one is listening."

I understood his frustration. David has been delivering the same message since the 1970s.

Early reporting about climate change was filled with

statistics and doomsday scenarios, and disseminated the usual images to underscore the crisis—factories belching smoke, crowded freeways, coal-fired power plants, and oil refineries. The message was: "The world is warming, and we did it. So stop doing all that damage or we're doomed." To compound an already complex problem, well-funded lobbyists and special interest groups have worked diligently to discredit climate scientists and challenge the whole idea of global warming.

In August 1966, James Garvey, president of Bituminous Coal Research, Inc., published findings in *The Mining Congress Journal* that stated, "The temperature of the earth's atmosphere will increase and vast changes in the climate of the earth will result." For decades after the report, the coal industry argued that the increased carbon dioxide in the atmosphere was beneficial to the planet.

In the 1970s, the oil industry published research that broadly agreed with the work done by the independent, peer-reviewed scientific community on global warming and climate change. However, the oil industry chose to organize a campaign of climate change denialism and public disinformation. Organizations that once worked to sell the virtues of smoking for the tobacco industry now crafted a narrative that argued that climate change was junk science. There was a lot at stake: an entire economic system based on finding, refining, and using petroleum.

Journalists who wanted to present both sides of the climate debate gave climate skeptics the opportunity to dispute the findings of climate scientists. Unwittingly, the news media helped amplify the counterposition that climate science was not based in absolute fact and was then, perhaps, not to be believed.

Once seeds of doubt have been sown, it is hard for what grows from them to not spread.

Much of what David Suzuki and I talked about did not make it out of an edit suite. "It's not how we want to convey climate to our audience" is what I and many other journalists across the media were told.

Perhaps how we initially addressed the crisis should have been focused more on solutions rather than on doomsday scenarios that garnered an emotional reaction.

Unfortunately, in a competitive news cycle, media consumers seemed to have rejected both narratives. The problem was too complex; the solutions seemed too monumental an undertaking, and the threat too distant to be of immediate concern.

But today, the story is different. As any accountant can attest, numbers don't lie.

Earth is getting significantly warmer. Twenty of the hottest years on record have occurred since 2000. Recent global temperature averages have been the warmest in the past two thousand years.

The Arctic is at its warmest in four thousand years. The thickness of sea ice has declined by 66 percent, or 2 metres, since 1960. The ice sheet covering Antarctica loses 275 billion metric tonnes of ice a year. The ice cap on Greenland has been losing 280 gigatonnes of ice per year since 2002; a gigatonne of ice is the equivalent of 1 cubic kilometre.

Changes in the Arctic are impacting global weather patterns that lead to persistent long-term weather events, like the drought and heat dome in western North America.

All of this has been forecast by climate scientists over the past fifty years. Many of us tuned out the story, until our children and grandchildren started pushing us to actually listen.

The children who listened to David Suzuki at the ROM in 2004 are now adults.

What has always moved our species forward has been our ability to think, learn, and work together. Our optimism and capacity to quickly change and adapt to new threats has built and rebuilt civilizations. It is wonderful news then that polling over the past several years reveals that climate change and the economy are the two most important issues for Canadians.

Finally, after too long, it's not just we individuals who are shouldering the burden of the climate crisis. Business and industry are now stepping up too. Again, an accountant will explain that the truth is in the numbers, and it is a disheartening fact that acting on climate change has been contingent on opportunity and profit formulas rather than on the clear threat laid out by climate scientists over the past fifty years.

We are acting—perhaps not as quickly as we should be, but generally people are moving together in the right direction.

This is what I was thinking about as I walked down the middle of Howe Street in Vancouver. I was shuffling through 15 centimetres of fluffy snow that lay, unplowed, on the main thoroughfare. Ahead was Robson Square, usually busy, but at 5:30 a.m. it was nearly empty.

It was January 2020; I was reporting on a snowstorm in Vancouver using an iPhone and a selfie stick.

In an average year, it will snow eleven times in the city. A typical winter offers about 35 centimetres of total accumulation during the entire season, or about 3 centimetres per snowfall. On this day, the city was expecting 26 centimetres, a big snowstorm by Vancouver standards.

Most people would just stay home, knowing that the snow will melt in a few days. In January, the average daytime high is 7 degrees Celsius and the low is 3 degrees; snow never stays on the ground for long.

Blizzard conditions had already closed the Sea-to-Sky Highway, and for the next twenty-four hours the city would be at a standstill. Schools and businesses would shut down, and flights and buses would be cancelled.

I ended my report in front of the Law Courts building; my camera panned down the empty street. The trees, holly shrubs, and occasional windmill palm on each side of the road were blanketed with newly fallen snow.

In Robson Square, several people were taking videos. A couple posed by the ice rink, taking the perfect selfie in this winter scene. All these images would likely join a steady stream of uploads to social media, sorted by hashtags like #BCStorm or #VancouverSnow, cross-checked, and monitored by AI algorithms for worldwide consumption.

Technology has enabled anyone with a smartphone, social media account, and connectivity the ability to report on any event. This fantastic development has rearranged, even upended, the traditional ideas and ideals of reporting. Now any story can be viewed from hundreds or even thousands of viewpoints.

But finding fact among so many different perspectives is left to the consumer, and sometimes it can be difficult to determine truth from misinformation or bold-faced lies. Are there really palm trees in Vancouver?

The term "fake news" was added to the *Oxford English Dictionary* in 2019. Defined as "news that conveys or incorporates false, fabricated or deliberately misleading information, or that is characterized as or accused of doing so," the term has also been used as an accusation against those a person may simply disagree with, not want to believe, or even want to discredit. This new twist in the gathering and distribution of news has opened a Pandora's box and is particularly concerning in an age when news stories are instantly measured by the number of clicks and views they receive.

This digital information trail is an income source for all platforms and services. Every "like" we trigger, each question we ask, all the information we view, and all the places our phone travels with us are sorted by artificial intelligence. Algorithms work endlessly to customize our digital experience based on what you like and do as an individual.

Is this as good as it is bad? I think that depends on the person.

More than ever before in our history, access to anything we'd like to learn about is instantaneous. As always, the great power that knowledge affords us also comes with great responsibility. Each of us now owns that burden of responsibility.

The data we consume—the selfies, news stories, tweets, emails, texts—all of it is just emotionless code, simple zeros and ones that become words and images. We convert the words and images into emotion. During ice storms, blizzards, floods, droughts, heatwaves, tornadoes, wildfires, and hurricanes, I have been able to document the real compassion that people demonstrate in times of need—from simple individual actions, like giving a warm meal to a family that has lost power during an ice storm, to the massive effort that raised nearly $300 million during the Fort McMurray wildfire. Witnessing and partaking in kindness toward others, especially during a crisis, is an enriching experience that helps all of us connect on an emotional level.

These same crises also revealed that there are weaknesses in how we manage information, how scientific facts and truth can be buried in a world overflowing with a never-ending stream of instantly accessible data.

Facts are facts, but sometimes our emotions cloud what is true.

The weather around us is always changing; it has always changed. We have no power to stop a drought or a hurricane. We are just witnesses to the moments of majestic beauty and to the awe-inspiring power of nature. But Duff's Ditch—the Winnipeg Floodway in Manitoba—wetlands conservation and solar farms in Ontario, windfarms along the front range of the Rocky Mountains, and countless other innovations illustrate our ability to adapt and thrive as our climate changes.

There will be missteps, misinformation, and mistakes. Learning to distinguish fact from fiction and staying focused on a sustainable and equitable future are how we will collectively craft a better tomorrow for our children.

ACKNOWLEDGMENTS

I hope you found as much enjoyment in those stories as I did telling them to you.

There are hundreds of people behind all the stories in this book. The people we spoke with in coffee shops and gas stations while filming a storm, the faces who filled a TV screen to talk about the heat or snow—they added the colour to our story about the weather. Thank you.

Behind the scenes at The Weather Network are a large group of editors, switchers, technicians, producers, writers, designers, reporters, and hosts, and an amazing team of meteorologists. All of them work around the clock 365 days a year. They love the weather and it shows. Thank you.

Pierre Morrissette has guided and grown The Weather Network family to become one of the world's preeminent suppliers of weather information on three continents and in several languages. "Mr. Weather Network" and I have always shared a storytelling vision, and I am happy he indulged mine for so many years. Thank you.

Pictures speak volumes. At The Weather Network, Edgard Mone and Mike Carroll helped find the right pictures that added extra depth to the stories in this book.

Books don't write themselves; they are a collection of ideas, thoughts, and emotions that find form through collaboration. Janie Yoon at Simon & Schuster Canada helped bring these stories to life. Together we found the best moments and emotions that needed to be shared. Stephanie Fysh then read and reread each word to ensure the story sounded perfect. Thank you.

Brian Wood is my agent; he laid the groundwork for this project to become what it became, a joy to write. Thank you.

"You should call the book *Weather Permitting*—that's what it's like here in Canada," Mike Campbell Critch said to me in late 2021. Thank you for the excellent title and for the funny moments in the chapter titled "The Rock."

Susan and Katharine, my wife and daughter, have always inspired me to make the stories better. "Go further, dig deeper, be in the moment, enjoy the journey," they said. Thank you, I love you both.

There are countless books about our environment and climate change; here are a few that I recommend.

Naomi Klein wrote a wonderful book called *The Battle for Paradise*, which examines what went so wrong during Hurricane Maria in Puerto Rico.

The Weather Makers by Australian paleontologist, mammologist, and environmentalist Tim Flannery takes a look at what our warming planet offers and what we can do to avert mass extinctions.

The Uninhabitable Earth by David Wallace-Wells is a realistic and disturbing look at how climate change will likely impact humankind if we don't will ourselves to seriously address this crisis.

Losing Earth by Nathaniel Rich describes how the decades-old proven scientific evidence of global warming has been intentionally denied for political gain.

Reporters and journalist use facts to tell stories. Facts about pollution and climate change can be found in a myriad of documents that have been filed away for decades. In August 1966, *The Mining Congress Journal* published an article by James Garvey, then president of Bituminous Coal Research, that discussed the correlation between fossil fuels and global warming. Facts presented in that article, a half-century ago, illustrated the seriousness of today's climate crisis. The full document can be found at ClimateFiles.com.

Weather can turn deadly. The tragic story of Janette and Mike Chapman and their son, Jeff, was related through the reporting of Glenda Luymes and Derrick Penner of the *Vancouver Sun* and Cathy Kearney of the CBC.

Details of the stories presented were drawn from my memories, experiences, personal notes, and subsequent conversations with the reporters, producers, and field teams I worked with during my twenty-seven years at The Weather Network.

ABOUT THE AUTHOR

CHRIS ST. CLAIR was a weather presenter and journalist on The Weather Network for more than twenty-five years. He is the author of the bestselling book *Canada's Weather: The Climate That Shapes a Nation*. He is also a popular speaker on meteorology and climate change. He lives in Kingston, Ontario. Connect with him on Twitter **@CStClair1**